Geography and Geographers

Geography and Geographers

**Anglo-American
Human Geography since 1945**

Second Edition

R. J. Johnston
Professor of Geography, University of Sheffield

Edward Arnold

© R. J. Johnston 1979, 1983

First published 1979 by
Edward Arnold (Publishers) Ltd
41 Bedford Square, London WC1B 3DQ

Edward Arnold (Australia) Pty Ltd, 80 Waverley Road, Caulfield East,
Victoria 3145, Australia

Edward Arnold, 300 North Charles Street, Baltimore, Maryland 21201,
U.S.A.
Second Edition first published 1983
Reprinted 1984

British Library Cataloguing in Publication Data

Johnston, R. J.
  Geography and geographers.—2nd ed.
  1. Anthropo-geography—Great Britain—History—
  20th century
  2. Anthropo-geography—United States—History—
  20th century
  I. Title
  304.2   GF28.G7

ISBN 0-7131-6387-9

Text set in 10/11 pt Plantin Compugraphic
by Colset Private Limited, Singapore
Printed and Bound in Great Britain
by Richard Clay (The Chaucer Press) Ltd,
Bungay, Suffolk

Contents

what, and for whom? – The liberal contribution – Radical alternatives – Liberals and radicals in debate – Conclusions

A Note to the Reader

This book comprises a book within a book. The inner volume – Chapters 2–6 – contains the factual record of the history of Anglo-American human geography. The outer volume, which incorporates Chapter 2–6 with Chapters 1 to 7, is an attempt to account for that history. Thus the outer volume is the 'innovative' work, based on the material of the inner volume: Chapter 7 interprets Chapters 2–6 in the light of the models presented in Chapter 1. The reader wanting only to obtain a knowledge of the contents of human geography since 1945 should begin at Chapter 2: the academic, too, may want to skip the first section of Chapter 1.

Acknowledgements

The author and publishers wish to thank the following for permission to use or modify copyright material: The Association of American Geographers for figs 3.2 and 3.3; C. W. K. Gleerup Publishers for fig. 5.2; Hutchinson & Co. Ltd for fig. 6.6; Methuen & Co. Ltd for fig. 3.1; Prentice-Hall International for figs 4.2 and 4.3; *The Statistician* for fig. 3.4; and the Institute of British Geographers for fig. 6.2.

Preface to the First Edition

Most students reading for a degree or similar qualification are required to pay some attention to their chosen discipline's academic history. The history presented to them commonly ends some time before the present. This has advantages for the historian, because the past is often better interpreted from the detachment of a little distance and there is less chance of hurting scholars still alive. But there is a major disadvantage for the students. In virtually every other component of their courses they will be dealing with the discipline's current literature, and so if the history ends some decades ago, they are presented with the contemporary substance but not with the contemporary framework, except where this is very clearly derivative of the historical context.

This state of affairs is unfortunate. Students need a conspectus of the current practice in their chosen discipline and should encounter a relevant overview which describes, and perhaps explains too, what scholars believe the philosophy and methodology of the discipline presently are and should be. Such an overview will allow the substantive courses comprising the rest of the degree to be placed in context and appreciated as examples of the disciplinary belief system as well as ends in themselves.

As a discipline expands, so the need for a course in its 'contemporary history' will grow too. In the last few decades, for example, most disciplines, and certainly human geography, have expanded greatly, with expansion measured by the number of active participants and their volume of published work. And the more active members there are, almost certainly the greater the variety of work undertaken, making it difficult for individual students to provide their own conspectus of the discipline from their own reading. Hence the need for a 'contemporary history' course at the present time.

The present book is the outcome of teaching such a course for several years, and is offered as a guide for others, both teachers and students. As with all texts, it has many idiosyncrasies. The course on which it is based is taught to final-year students reading for honours degrees in geography, and is probably best used by people at that level since it assumes familiarity with the concepts and language of human geography. Further, my view is that students probably benefit most from a framework after experiencing some of its contents; this book provides the matrix for organizing the individual parts rather than a series of slots into which parts can be placed later, which would be the case if the course were taught early in the degree.

A series of constraints circumscribes the contents and approach of this contemporary history. First, it deals only with human geography, for several reasons. The most important is that I find the links between physical and human geography tenuous, as those disciplines are currently practised. The major link between them is a sharing of techniques and research procedures, but these are shared with other disciplines too, and are insufficient foundation for a unified discipline. (What price a department of factorial ecology?) Further, my own competence, work and interests lie wholly within human geography and although I have been trained with and by and have worked among physical geographers, and have obtained stimuli from this, I am incompetent to write about physical geography. And finally, much of the human geography discussed here is of North American origin, and many geographers there, especially in the United States, encounter no physical geography as it is understood in British universities. To a considerable extent, therefore, human and physical are separate, if not independent, disciplines. Throughout the book, I use the terms 'geography' and 'human geography' interchangeably.

The second constraint is cultural; the subject matter of the book is Anglo-American human geography. Most of the work discussed emanates from either the United States or the United Kingdom: there are some contributions from workers in Australia, Canada, and New Zealand, but the efforts of geographers in the rest of the world are largely ignored. (A partial exception is Sweden, which has major academic links with the Anglo-American tradition; much Swedish geographical work is published in English.) Such academic parochialism in part reflects personal linguistic deficiencies, but it is not entirely an idiosyncratic decision. Contacts between Anglo-American human geography on the one hand, and, say, that of France and of Germany on the other have been few in recent decades, so to concentrate on the former is not to commit a major error in separating a part from an integrated whole.

The final constraint is temporal, for the book is concerned with Anglo-American human geography during the decades since the Second World War only. Again, this is in part a reflection of personal competence, for I have been personally involved in academic geography for the last twenty years. But the Second World War was a major watershed in so many aspects of history, not least academic practice, and much of the methodology and philosophy currently taught in human geography has been initiated since then.

This book, then, is a history of Anglo-American human geography since 1945. It does not purport to be an objective history, for such an enterprise is impossible. The material included represents subjective judgements: the stress on certain topics is subjective, too, and so is the organization. But although not objective, nor intended to be, the book is neutral. My own opinions are not stated, and are not intendedly implied in anything that has been written (though some of them may be identifiable). There is no commentary, only a presentation of what I perceive to be the salient features.

Arising out of this intended neutrality is a second characteristic of the book, its dependence on the written statements of others. There are few lengthy quotations, but many short ones. Most arguments have, of necessity, been précised or paraphrased, in which case there may be unintentional distortion of the emphasis in the original. My aim has been to report what others have written, and the contents of Chapters 2–6 depend entirely on the published record, for no use has been made of personal memoirs (however valuable these might be, if a representative set could be collected). A consequence of this orientation is a large bibliography. Some authors are commended by reviewers for the utility of their bibliographies. Mine indicates the material on which the book is based.

Within the terms of reference set, I have not attempted simply to provide a chronology of arguments. As well as describing the changing contents of Anglo-American human geography I have tried to account for them, to suggest why certain things were said and done, by a particular person in a particular place. Such an attempted 'explanation' of events cannot be neutral, and so it has been handled by writing a book within a book. The outer volume is Chapters 1 and 7; the inner is 2–6: the former presents the subjective account, phrased in terms of the models developed by historians of science, and the latter contains the neutral description. The inner book can be read independently of the outer; the contents of Chapter 7, however, are derived from all of those preceding it.

The account provided in the outer book is not idealist (using that term in the same way as in Chapter 5), for in attempting to explain changes in attitudes to philosophical and methodological topics I rely on my own modelling of scientific progress rather than on the views of the change-agents themselves. The modelling is set in the context of other studies of the history of academic disciplines, notably the physical sciences. By the end of Chapter 7, this model turns out to be something of a straw man. I am not the first to use it within human geography, however, and so my presentation, criticism and then replacement of it represent a contribution to the general enterprise of writing the history of geography (and perhaps of other social sciences).

One problem with writing 'contemporary history' is knowing when to stop. There is a constant flow of new material which can be employed in the task, and there is always the probability (sometimes the certainty, using publishers' advertising) of something very useful appearing tomorrow. Eventually, a halt must be called. For this book, it was in mid-1978, although some later material is referenced (because I was privileged to read it before it went to press). By the time it appears, therefore, this book will of necessity be a period piece, although the arguments in Chapter 7 suggest that a few years will elapse before a major revision is needed.

It was Malcolm Lewis who designed the course of which the material in

4 *Geography and Geographers*

this book became a part; Stan Gregory was responsible for my teaching it. Both are in no way responsible for the outcome. I am grateful to them for the opportunity to undertake what turned out to be, for me, an enjoyable and stimulating task. The comments (direct and indirect) of the various student audiences have helped to redefine the contents in various ways over the years, and I am grateful for these.

Preparation of this book has been helped materially by several people. My wife Rita has, as always, been of great assistance in a variety of ways, not least in reading the whole manuscript twice and giving much advice on the presentation. Secondly, Walter Freeman has read the entire manuscript and given much of his time in discussing both its contents and its context; I am deeply grateful to him for his interest, his kindness, and his continued friendship. Alan Hay, too, read the complete first draft and commented freely and very helpfully on many aspects of the work. Both he and Walter Freeman felt that there is not enough of me and my opinions in the book: I hope that they understand why. The diagrams were produced by Stephen Frampton and Sheila Ottewell, and Joan Dunn has yet again created an excellent typescript out of a messy manuscript. My thanks to all.

I make it clear in Chapter 1 that the progress of any individual's academic career depends considerably on the actions (and sometimes inactions) of others. No academic is an island. Many people have helped me in the last twenty years, but I should like to express special thanks to Percy Crowe, Walter Freeman, Basil Johnson, Murray Wilson, Barry Johnston, Michael Wise, Stan Gregory and Ron Waters.

Autumn 1978   R. J. Johnston

Preface to the Second Edition

The generally favourable reaction to this book, and the continued demand for it, suggests that a successful formula was identified in its writing, and that no basic changes are needed. Nevertheless, it was felt that a second edition was desirable, rather than a reprint. The opportunity has been taken to correct errors, to update (hence expanding the bibliography substantially) the 'inner volume' (to mid 1982; the first edition was written in 1978), and to respond to some of the criticisms, in particular with regard to the contents of the 'outer volume', on which my ideas have changed somewhat.

The preface to a second edition should not be used simply as a response to particular reviews of the first. But the reviews of this book have raised a number of general issues that are worthy of brief discussion here – and which have been taken into account when preparing the revision.

The first point concerns the nature of the entire enterprise. In the original Preface, I stated that although the book was not objective history – 'for such an enterprise is impossible' – it was neutral, in that it contained no personal commentary. In retrospect, this was an over-statement. The book is basically a reconstruction, from published material, of philosophical and methodological debates within human geography; it is more concerned with writing about human geography than with writings that are human geography (as one reviewer so clearly pointed out), and is something of a political history of human geography. In effect, it is an exercise using the philosophy of idealism outlined in Chapter 5 (despite my statement to the contrary in the original Preface). The literature that I review represents the theories (or ideologies) of its authors regarding the nature of human geography. The framework into which I set that literature represents my theory of the history of human geography, which has been constructed – and is continually reconstructed – to ensure that the two cohere (see Chapter 7). It is, then, a personal statement, set in the context of my own socialization as a human geographer.

This orientation has not been changed. I still rely entirely on published evidence for my history – while accepting the value of 'oral history' of various kinds, which is now providing valuable published materials (Browning, 1982; Buttimer, 1983; Buttimer and Hagerstrand, 1980). But the revision has allowed me to remedy some important omissions, and to reorganize certain sections to emphasize more fully the contributions of particular individuals. It is their contributions to philosophical and methodological debates that are the focus of the book. Substantive contributions are widely reviewed in a variety of places – most notably the excellent journal *Progress in Human Geography*. It is not a major purpose of this book to summarize what human geographers have done, but rather to concentrate on how they have done it, and why.

Some criticism has been directed at the book for its separation of human from physical geography; indeed, I have been 'accused' of doing the discipline a political disservice by advancing this separation. I stand by my statement in the original Preface, however – 'I find the links between physical and human geography tenuous, as they are currently practised'. During the 'spatial science' era (Chapters 3 and 4) there was a common interest among physical and human geographers in methodological questions, especially of a statistical and mathematical nature, and the interactions that took place are recognized here. And certain methodological frameworks – notably systems analysis (Chapter 4) – seemed to provide common bonds. These bonds remain, and are presented here. But as each of the two fields has moved away from quantitative description to studies of, first, how (process studies to physical geographers) and, then, why (process studies to human geographers), so the differences between the two have been magnified. There is very little basis for bonds with physical geography in the approaches to human geography discussed in Chapters 5 and 6 of this book. Of course, the subject matter of human geography is closely concerned with that of physical geography, but there is no apparent need (and certainly no published evidence of its realization) for an integration of the two fields as they are currently practised. Some argue that such integration is (or should be) achieved through the study of resources, but a review of the relevant literature (Johnston, 1983a) provides no evidence to support that claim. Human and physical geography overlap. There are benefits to be achieved from that, especially in a pedagogical context, and I would not strongly advocate their institutional separation. But they are separate fields, each overlapping more with other fields than with each other.

Internal to human geography, it has been suggested that my treatment is unbalanced. In particular, cultural, historical and regional geography are ignored, relative to economic, social and political geography. To the extent that this is so, it reflects relative contributions to the literature being reviewed. The focus of this book is philosophical and methodological debate, not substantive contribution.

One of the most fundamental criticisms concerns the underlying rationale for the book, that there is a separate history of human geography worthy of investigation. By presenting such a history, am I not drawing artificial boundaries? To this, I would respond that the boundaries already existed. Human geography is an institutionalized discipline in the countries studied here; at the school level in Britain it is clearly defined by the syllabuses for public examinations. The boundaries around human geography are indeed artificial, because the subject matter of the social sciences should not be compartmentalized. But a desire to break down those boundaries in some way does not detract from the need for a history which, among other things, should demonstrate to students how that desire has developed within the academic community of human geographers in recent decades.

Despite the institutionalization of human geography, there is nevertheless considerable movement through its boundaries, involving contributions by human geographers to other disciplines and vice versa. This may be understated in the book. It takes a variety of forms. The basic flows are of ideas from other fields into human geography. These ideas are assimilated into the recipient discipline, and in return human geographers may seek to demonstrate the results of that assimilation to the source discipline. This is usually done by way of publications in the journals of the latter; there has been more of it in some source disciplines (agricultural history, for example) than others (such as political science; Laponce, 1980). Some individuals have transferred their disciplinary allegiance as a result: there are several British professors of sociology who were trained as geographers, for example. But are they still human geographers, whose work should be reviewed here?

Certainly, inter-disciplinary links are substantial, even if not always apparent, and should not be ignored. Developments in human geography in recent decades have been indelibly influenced by these links – many of them completely informal, because they involved 'World Three' (p. 217) only. But such links do not deny the existence of a discipline called human geography into which many students are socialized each year and whose history is of intrinsic interest.

Finally, why confine the treatment to the 'English-speaking world', basically to Britain and North America? My answer is the same as it was in the Preface to the first edition: competence. The need for a more thorough international survey is clear, however, and small steps have been taken to meet it (Johnston and Claval, 1983). There are other things we are ignorant about. None of the arguments presented here is in any way backed by solid quantitative evidence, where that would be relevant. Gatrell (1982) has pointed to one possible direction, but the routes are wide open. Indeed, the tasks ahead of us are many

> The 1980s promise to be challenging, exciting and fraught years, not least to the academic community. For geographers, the uncertainty is a double one: both in what to study and how to study it (Robson, 1982, p. 1).

In continuing the work on this subject, and in the formulation of the second edition, my debts from the original book remain. I am particularly indebted to my publishers for constant encouragement, and also for stimulating me to provide a companion volume – *Philosophy and Human Geography* – that outlines the philosophies the debates over which are reviewed here. The contributors to *The Dictionary of Human Geography* stimulated me considerably, as did the valuable comments of the kind reviewers of the first edition. To them all, I am extremely grateful. As always, I am deeply indebted to Rita Johnston and Joan Dunn.

Summer 1982   R. J. Johnston

1

The nature of an academic discipline

This book is a study of an academic discipline, in particular of its content. But content cannot be discussed fully without an understanding of context, and it is to provide that context that the present chapter has been prepared. In essence, the context is the population involved in the discipline. To study an academic discipline is to study a society, which has a stratification system, a set of rewards and sanctions, and a series of bureaucracies, not to mention a large number of interpersonal conflicts (some academic, some not). To the outsider, academic work may seem to be objective and 'scientific'. But many subjective decisions must be taken: what to study; whether to publish the results; where to publish them and in what form; what to teach; whether to question the work of others publicly; and so on. As with all human decisions, they are made within the constraints set by the society of which the academic worker is a member.

In studying an academic discipline one in effect has to study a society within a society; both are relevant in setting the constraints to individual as group activity. The study of the society within the society involves a focus on two questions; 'how is academic life organized?'; and 'how does academic work, basically its research, proceed?'. The first two sections of this chapter are concerned with those two questions. Use of the term 'society within a society' strongly implies that academic life does not proceed independently. It is not a closed system but rather is open to the influences and commands of the wider society which encompasses it. A third question to be asked, therefore, is 'what is the nature of the society which provides the environment for the academic discipline being studied, and how do the two interact?'. This is the subject of the final section of the chapter.

Academic life: the occupational structure

Pursuit of an academic discipline in modern society is part of a career, undertaken for financial and other gains; to most of its practitioners, the career is a professional one, complete with entry rules and behavioural norms. In the initial development of almost all of the academic disciplines studied today, some of the innovators were amateurs, perhaps financing their activities from individual wealth. There are virtually no such amateurs now; very few of the academic publications in human geography are by other than either a professional academic with a training in the discipline or a member of a related discipline. The profession is not geography, however. Rather geography is the discipline professed by the individual academic who probably states his profession as university teacher or some similar term. Indeed, the great majority of academic geographers are teachers in universities or comparable institutions of higher education. (Recent years have seen the invention of new types of higher education institution. The term university is used here as a generic term, therefore. In effect, universities remain the main locations of academic research.) Academic geographers are distinguished from other geographers (many of whom are also teachers) by their commitment to all three of the basic canons of a university; to propagate, protect, and advance knowledge. It is the advancement of knowledge – through the conduct of fundamental research and the publication of its original findings – which identifies an academic discipline; the nature of its teaching follows from the nature of its research.

The academic career structure

Members of university staffs – hereafter termed academics – follow their chosen occupation within a well-defined career structure. Two variants of this structure are relevant here: the *British model* and the *American model*. Entry to both is by the same route. The individual must have been a successful student, as an undergraduate and, more especially, as a postgraduate. While the latter, he will have undertaken original research, guided by one or more supervisors who are expert in the relevant specialized field. The results of this research are almost invariably presented as a thesis for a research degree (usually the PhD), which is examined by experts in the field of study. The PhD is an almost obligatory entrance ticket for the American model of the career structure; in the British model it is usual, but far from universal.

While undertaking the research degree, most postgraduate students will have obtained some experience in teaching undergraduates, most particularly in practical and tutorial classes: indeed, some universities, notably those operating the American model, finance their research students through their teaching activities. Having completed the research degree, the individual may then proceed to further research experience, or he may gain appointment to a limited-tenure teaching post in a university department. This appointment may be for a limited period only, offering an 'apprenticeship' in the teaching aspects of the profession whilst either the research degree is completed or the individual's research expertise is consolidated.

Beyond these limited-tenure positions lie the permanent teaching posts, and it is here that the British and American models deviate (Figure 1.1). In the British model, the first level of the career structure is the lectureship. For the first years of that appointment (usually three) the lecturer is on probation: annual reports are made on his progress as teacher, researcher and administrator, and advice is offered on adaptation to the demands of the profession. At the end of the probationary years, the appointment is either terminated or confirmed.

Appointment in this model is almost always to the staff of a department; most departments are named for the discipline which the individual staff member pursues. The prescribed duties involve the undertaking of research and such teaching and administration as directed by the head of the department. The lecturers are on a salary scale, and receive an annual increment; accelerated promotion is possible in some universities. In the United Kingdom there is an 'efficiency bar' after a certain number of years' service; 'crossing the bar' involves a promotion, and is determined by an assessment of the lecturer's research, teaching and administrative activities.

Beyond the lectureship are further grades into which the individual can be promoted. The first of these in the British model is the senior lectureship for which there is no allocation to departments; entry to it is based again on an

Figure 1.1 The academic career ladder. Note that whereas in the American model it is very unusual not to progress up through the stages in an orderly sequence, in the British model missing steps is quite common (i.e. some lecturers move direct to readerships, even to professorships, without first being senior lecturers – there are no senior lecturers in the 'Oxbridge' universities – and many professors are recruited from the senior lecturer rather than the reader grade)

assessment of the lecturer's conduct in the three main areas of academic life (research, teaching and administration), and is based on open competition across the whole university. The next level is the readership, a position which is generally reserved for scholars of high quality; the criteria for promotion to it focus strongly on research activity by the candidate.

The final grade (associated with academic departments as against administration of the university as a whole) is the professorship. Although superior in status to the others, this is not simply a promotional category: most professorships are appointed to from open competition preceded by public advertisements. Initially the posts of professor and of head of department were synonymous, so a professor was appointed as an administrative head, to provide academic leadership. With the growth in department size in recent years, however, and the development of specialized sub-fields within disciplines, each requiring its own leadership, it has become common for departments to have several professors. In some, the position of head remains with a single appointee; in others, probably the majority, it rotates among the professors; in a few, the headship is now an elected position, independent of the professoriat, to which other grades may aspire.

Finally in the British model, promotion is possible to non-advertised, personal professorships. These positions are relatively rare, and are reserved for distinguished scholars who have risen through the ranks of the one establishment and for some reason have not obtained appointment to a professorship (a 'chair') in the normal manner.

The system under the American model is similar, but generally simpler (Figure 1.1). There are three permanent staff categories: assistant professor, associate professor, and professor. The first contains two sub-categories – those staff on probation who do not have 'tenure' and those with security of tenure. Each category has its own salary scale but not automatic annual increments. (In the United States, these scales vary from university to university – according to prestige, State etc.) Salary adjustments are the result of personal bargaining, on the basis of academic activity in the three areas already listed, and as a result it is possible for the salary scales of the categories to overlap in the same department. Movement from one category to another is promotional, and is a recognition of academic excellence. Professorships are simply the highest promotional grade and do not carry any obligatory major administrative tasks. Heads of departments in the American model are separately appointed (most are designated as chairmen), usually for a limited period only, and they need not be professors (usually referred to as 'full professors').

Status, rewards and promotion prospects

The rewards of the academic profession are the salary levels, plus the general status of the profession, the particular status of its individual promotional categories, and the relative independence and flexibility of the working conditions. The tangible rewards – the payments – are determined by the academic's category and his position within it. Promotion must be 'earned'. How?

Academic work has three main components – research, teaching and administration. Entrants to the profession have little or no experience of academic administration and their only teaching has probably been as assistants to others. Thus it is very largely on proven research ability that the potential of the aspirant for an academic career is judged. To some degree, research ability can be equated with probable teaching and administrative competence, since all require the same personal qualities – enthusiasm, orderliness, incisive thinking, and ability to communicate orally and in writing; to a considerable extent, however, academics are appointed to their first position on faith, hence the usual probationary periods before tenure is granted.

Once admitted to the profession, the academic will have to undertake all three types of work, so that promotional prospects can be more widely assessed. In effect, teaching and administration probably carry less weight with those responsible for promotions than does research. This is partly because of difficulties in assessing performance in the first two, and partly because of a general academic ethos which gives prime place to research activity in the evaluation of peers.

Although not necessarily progressive in terms of increasing level of difficulty, administrative tasks tend to be more complex and demanding of political and personal judgement and skills as the academic becomes more senior. As made clear by the 'Peter Principle' (Peter and Hull, 1969), a person's ability to undertake a certain task can often only be fully assessed after he has been promoted to the relevant position. Promotion must frequently be based on perceived potential, and although it is possible to point to somebody whose administrative skills are insubstantial it is not always easy to assess who will be able to cope with the more demanding administrative tasks.

It has long been realized that assessment of teaching ability is difficult (although clear inability is often very apparent). Student and peer evaluations are possible and useful, as is scrutiny of the work done by the students taught. But criteria for judging teaching performance at university level are ill-defined, and what is expected from a lecturer/tutor/seminar leader often varies quite considerably within even a small group of students. Thus the majority of teachers are usually accepted as competent and undistinguished, and their performance is neither help nor hindrance to their promotion prospects.

And so the main criterion for promotion is usually research performance, although a relatively undistinguished record in this area can be compensated by excellence elsewhere. How then is research ability judged? Details of how research is undertaken are considered in the next section; here the concern is with its assessment rather than with stimulus and substance.

Successful research involves making an original contribution to a field of knowledge, in a variety of ways. It may involve the collection, presentation, and analysis of new facts, within an accepted framework; it may be the development of new ways of collecting, analysing and presenting facts; it may comprise the development of a new way of ordering facts – a new theory or hypothesis; or it may be some combination of all three. The originality of a contribution is judged through its acceptance, or validation, by those of proven expertise in the particular field. The generally accepted validation procedure is publication, hence adages such as 'unpublished research isn't research', 'publish and be damned', and 'publish or perish'.

The outlets for the publication of research findings are the scholarly journals, which operate fairly standard procedures for the scrutiny of submitted contributions. (There is general belief that the journals published by academic societies tend to have higher standards than those published by commercial companies, but the validity of this is difficult to assess.) Manuscripts are submitted to the editor, who seeks the advice of qualified, often senior, academics on the merits of the contribution; these referees will recommend either publication, rejection, or revision and resubmission. When accepted, a manuscript will enter the publication queue, which in some cases may be up to two years long, depending on the academic reputation of the journal.

Although widely accepted, this procedure has certain inbuilt flaws, largely because it is operated by human decision-makers. The opinions of both editor (on the manuscript, and on whom he chooses as referees) and referees may be biased in some way, so that a paper can be rejected by one journal but accepted by another, without alteration. In most disciplines there is an informal ranking of journals in terms of their prestige, and it is considered more desirable to publish in some rather than others. (This prestige ranking is sometimes taken into consideration by appointment and promotion committees.)

Some research results are published in book form, rather than as articles in journals. Most academic books are texts, however, published by commercial companies whose main interest is marketability among the large student population. The textbook may be innovative in the way that it orders and presents material, and can be beneficial to its author's reputation (as well as to his bank balance), but it is not usually a vehicle for demonstrating research ability. Many companies, including the university presses, do publish research monographs, however, to present the results of major research projects to relatively small academic markets. Their decisions on whether to publish are made on academic as well as commercial grounds, usually with the aid of academic referees, and their output is validated through the book review columns of the journals.

Processes of promotion and appointment: patronage
Whatever the weighting given by the relevant committees to the three main academic activities (there are others pursued by some, such as consulting for

outside bodies), there is still a major question of how such committees make their assessments. The only 'objective' information which can be presented to them are examination (degree) results and lists of publications which have been validated by academic journals. But how is such information to be evaluated?

Two modes of assessment are widely used and relied upon: the written opinion of a third party (a referee) and the interview. In the British model, considerable weight is placed on the former. An applicant for a position must supply a list of referees who will provide an opinion on his suitability for the post; fairly clearly, he is likely to ask those who he has worked with, and who are favourably inclined towards him, to act in this capacity. As appointment committees tend to be swayed very much by these reports, especially in their preparation of a shortlist of candidates to be interviewed, the opinions of well respected members of the discipline who act as referees are often crucial. Thus certain leaders in a subject often find it easier to get their candidates appointed to posts than do others: there is a considerable element of patronage involved in obtaining a university post, especially a first university post. This patronage is particularly important in getting the candidate onto the shortlist of those given an interview, from which the final selection is made.

Promotions under the British model are largely based on the opinion of one person, the head of the candidate's department. Reports must be made on each lecturer: annually during the probation period, and then at the confirmation stage after probation, on reaching the 'efficiency bar', and for either accelerated promotion within the lecturer scale or promotion to senior lecturer. Again, therefore, there is a strong element of patronage, although constraints are built into the committee system to try and ensure fairness of all, including the right either to present one's own case or to appeal against a decision. Some universities use outside referees to provide extra evidence for cases of promotion to senior lecturer, and promotions to readerships always involve the use of external referees, usually nominated by the head of department concerned.

Referees and interviews are also used in the appointment of professors. Two sets of referees are used. The first contains those nominated by the candidate as being prepared to provide a confidential report on his potential for the job in question; the second is a group of assessors nominated by the university, usually senior academics in the relevant field. The latter not only comment on the list of candidates who have applied but also suggest other worthy candidates who might be approached: their potential patronage power is great.

Procedures are slightly different under the American model. More weight is usually given to an enlarged interview, often involving candidates in giving seminars to the department and meeting with various groups of staff. The reference is still important, however, especially for aspirants to first teaching positions, and a letter from a respected leader in a field can be very influential in gaining appointment for his former student. In general, the individual candidate is more active himself in this system, however, perhaps canvassing for interviews during the annual conferences of scholarly societies, for example. For promotions and for salary rises there is considerable bargaining between individual, department chairman, and university administrators: external referees' opinions may be sought when tenure is being confirmed, but not usually thereafter.

In all of these procedures the applicant, for a new position or for a promotion, depends to a considerable extent on the opinions of senior academic colleagues in the evaluation of his prospects. Some opinions carry more weight than others, and so it is important when developing a career to identify potentially influential individuals, to keep them informed of your work, and to enlist their support for your advancement. Because of the lack of truly objective criteria for measuring research, teaching and administrative ability, such patronage is crucial.

Other rewards and the sources of status

The tangible rewards of an academic career are the salary and the life style, plus the occasional extra earnings that are possible – for examining, writing, lecturing and consulting. In addition the hours are flexible; the possibilities for travel are considerable; and the constraints on when, where and how work is done are relatively few, compared to most other professions. And there are other, less tangible rewards. Involvement with the intellectual development of students is a considerable reward for many, and brings much satisfaction. There is also the charisma associated with recognized excellent teachers, and even more so by leading researchers, whose publications are widely read, whose invitations to give outside lectures are many, and whose opinions as examiners, referees and reviewers are widely canvassed. And the conduct of research brings its own rewards, apart from the charisma; the satisfaction of having identified and solved a significant problem is often considerable, and a major reward for the dedicated researcher.

One reward common to social systems exists in the academic community too: power. Patronage is power, as is any work as examiner, referee, or reviewer. The careers of others are being affected, and exercise of this power can bring with it the loyalty and respect of those who benefit. Because the academic system is so dependent on the opinions of individuals, because the opinions of some individuals are more valued than are those of others, and because power over others is a 'commodity' widely desired in most societies, many academics seek positions of influence.

Among the most influential positions within an academic discipline are the administrative headships of departments. Holders of these positions can instruct other staff members (usually after consultation) in their teaching and administrative duties; they are frequently used by staff and students as referees for job and other (such as research grant) applications; and their reports are crucial in the promotions procedure. The departmental organization of universities is a bureaucratic device which makes for relative ease in administering what is often a very large institution. (It also tends to fossilize disciplinary boundaries, as will be discussed later.) Departmental heads not only have power over members of their own discipline's staff, they also participate in the administration of the university as a whole, and they treat within the university committee system for departmental resources. As in all bureaucracies, there is a tendency for the status and power of the various departmental heads to be a function of the size of their 'empires' (Tullock, 1976). Heads of large departments, especially large and growing departments (growth being generally considered as a 'good thing'), are very often the most important individuals in a bureaucracy. Thus they have the incentive, whether or not they are permanent occupants of the headship, to build up their departments, which usually means increasing student numbers, since universities tend to allocate resources to departments according to the numbers of students taught. This brings power and prestige, both within the university and beyond; it is an added reward for the academic bureaucrat, and the power over resources which it involves is usually of benefit to his whole department.

Finally, the academic bureaucrat, whether or not a head of department, can gain power beyond his home university through, for example, obtaining positions on committees. These may be concerned with his subject through professional bodies, with the allocation of research moneys by public or private foundations, or with a wide range of public duties. Again, the status and power obtained can spill over to others, since patronage is important in all of those roles.

The academic working environment

The continuing goal of an academic discipline is the advancement of knowledge. Each discipline pursues that goal with regard to a particular area of study. Its individual members contribute by conducting research and reporting their findings, by integrating material into the disciplinary corpus, and by pedagogical activities aimed at informing about, promoting and reproducing the discipline: in addition, they may argue the discipline's 'relevance' to society at large. But there is no fixed set of disciplines, nor any one correct division of academia according to subject matter. Those disciplines currently in existence are contained within boundaries established by earlier communities of scholars. The boundaries are porous, so that disciplines interact. Occasionally the boundaries are changed, usually through the establishment of a new discipline that occupies an enclave within the pre-existing division of academic space. (The boundaries are not necessarily the same in all comparable institutions.)

Just as there is no immutable set of disciplinary boundaries so there is also no right way of undertaking research nor, in many cases, any exact criteria for determining whether research findings, and even more their interpretations, are correct. The 'right' and 'wrong' ways of doing research, the 'correct' and 'incorrect' interpretations of research findings, and the 'proper' ways of presenting knowledge and training students are all the product of decisions by academics themselves. Thus, not surprisingly, there is considerable debate within disciplines over these issues. At any one time, there may be consensus within a disciplinary community regarding both its subject matter and its research methods. But controversy is just as likely, as academics discuss the relevance of particular research findings, the validity of certain research methodologies, and so on.

The study of the controversies and consensuses that characterize academic disciplines is the function of historians of science. (Science is used here as a very broad term to encompass the entire range of academic disciplines in the natural and social sciences.) To the outsider, the workings of a science are generally mysterious – especially regarding disciplines that require a great degree of prior training before original research is undertaken and whose literature is virtually impenetrable to the untrained. It is generally believed, however, that science is an objective activity undertaken within very strict rules, and involves the continuous excitement of the search for new discoveries. Indeed, scientists frequently present themselves in this light. A certain set of values, it is suggested, is universally subscribed to within academia. The main five (according to Mulkay, 1975, p. 510) are:

1   The norm of originality; academics strive to advance knowledge, and they conduct original research designed to discover and account for new aspects of the world;

2   The norm of communality; all information is shared within the academic community – it is transferred through accepted channels (notably the research journals) and its provenance is always recognized when it is being used;

3   The norm of disinterestedness; academics are devoted to their subject, and their main reward is the satisfaction of participating in the advancement of knowledge – a reward that may bring with it charisma and promotion;

4   The norm of universalism; judgements are made on entirely impartial criteria, which take into account the academic merits of work only and make no reference to the personalities of the researchers; and

5   The norm of organized skepticism; knowledge is furthered by a continuing process of constructive criticism, by which academics are always reconsidering both their own work and that of others. According to these five norms, therefore, academic work is carried out in a neutral fashion; there is a complete lack of partiality, self-seeking, secrecy and intellectual prejudice. The existence of objective criteria for assessment is assumed, as are high levels of ability and humility on the part of members of the academic community.

Against this ideal view of science and scientists, one which many of the latter seek to promote, are the results of studies in the history of science (most of which refer to the physical sciences: Mulkay 1975; Mulkay, Gilbert and Woolgar, 1975). These show that the procedures adopted are frequently by no means objective and neutral, and the picture they present is of disciplines that grow 'by gathering more detail in areas already investigated, and by stumbling across new sets of facts in areas of experience never previously investigated' (Barnes, 1974, p. 5). Science as a whole is a culture, and each discipline within it is one (or more) subculture(s). It has its own rules and procedures which, like those in other cultures and subcultures, are open to change as a result of internal decision-making. Furthermore (as will be suggested in more detail below), scientific cultures are parts of wider cultures, and although to some extent scientists – because of their (self-imposed) expert status – can impose their own views of themselves on their host societies, they are subject to external influences (see Barnes, 1974).

Each scientific discipline is a separate academic community, therefore; many are groupings of several related communities. The goal is the advancement of knowledge, but the definitions of advancement and of knowledge itself are influenced, if not determined, by the members of the community. Thus the study of the history of a discipline is not simply a chronology of its successes. It is an investigation of the sociology of a community, of its debates, deliberations and decisions as well as of its findings. (See Watson, 1968, for an example of this; for a fictional account, see Cooper, 1952.) Such investigations may take one of two general forms. The first is the empiricist which seeks to portray history as it happened. In this, each community is treated separately and is the subject of a specific history. The other seeks to generalize, to understand how sciences progress. This involves either an inductive approach – studying individual histories and identifying their common themes – or a deductive framework that provides a model of scientific progress against which events in a particular discipline are compared. Most of the models currently available have been developed for, and tested on, the physical sciences. Their relevance to human geography has been suggested, however (e.g. Haggett and Chorley, 1967; Harvey, 1973) and they are presented here as a backcloth for a later evaluation of the substantive material arranged in Chapters 2–6.

Kuhn, normal science, and revolutions
The framework for studying the history of science that has received most attention (indeed, the book is one of the most frequently cited by social scientists) is T. S. Kuhn's (1962, 1970a) *The Structure of Scientific Revolutions*. As Kuhn himself (see Kuhn, 1970b) and many commentators have pointed out, this work is a study in the sociology of science. It is a positive interpretation – a presentation of what scientists do – rather than a normative programme – an argument for what scientists *ought* to do. And it seeks to make generalizations, to identify common elements in the histories of scientific disciplines, although these generalizations are not the basis for a predictive methodology. Kuhn's goal is to identify 'what science, scientific

research as it is actually practised, is really like' (Barnes, 1982, p. 1).

According to Kuhn, scientists work in communities, groups of researchers and teachers who share a common approach to their work. They operate within an agreed philosophy, concur on the theoretical focus of their work and use accepted methodological procedures. Their research involves using the procedures to solve problems identified within the theoretical framework, thereby adding to knowledge (the store of problems solved) and extending the range of their theory. The framework, its procedures, and its empirical substance are codified in their textbooks. What they share is termed a *paradigm* by Kuhn; 'a scientific community consists of men who share a paradigm' (Kuhn, 1970a, p. 176). What this means, as Popper (1959) expresses it, is that once a scientist is socialized into a research field he can proceed directly to its unsolved problems. The existing framework defines these problems, providing a context for the research, and a methodology is available with which they can be tackled. The researcher does not set out to tackle a problem *de novo*, therefore, but on the basis of what is already known in the chosen field. Thus a paradigm is 'an accepted problem-solution' (Barnes, 1982, p. xiv), which by its very solution poses the next problem. Progress in science is achieved by problem-solving.

To be able to undertake research within a paradigm requires an understanding and acceptance of its philosophy and methodology. This is provided by a period of training, during which the potential new researcher is socialized into the paradigmatic culture, its way of thinking about its scientific problems. The key to this training is the paradigm's textbooks, the summaries of its literature which define what is known and how more knowledge can be obtained. (That literature may be very large and so must be sampled: for geography, see Stoddart, 1967a.) The purpose of this training is preparation for work within an accepted mould: 'Scientific training is dogmatic and authoritarian . . . [it] does not generate or encourage traits such as creativity or logical rigour; rather it equips scientists so that it is possible for them to be creative, or rigorous, or whatever else, in the context of a specific form of culture' (Barnes, 1982, pp. 16–17.)

Having been integrated into the paradigm through a period of training, the scientist joins a research community. Such communities, sometimes termed 'invisible colleges', operate, usually in groups, through close interaction involving attendance at specialized conferences and the private circulation of research papers in pre-publication form, and reflected in citations to each other's work (Crane, 1972). Within such communities, individuals receive recognition for the value of their work and attract the patronage of leaders who may assist in their career advancement; some communities have only a few patrons, perhaps even one, and may be identified as a particular school of thought.

The nature of research within a paradigm involves filling in the gaps; the researcher 'has to make the unknown into an instance of the known, into another routine case' (Barnes, 1982, p. 49). The paradigm provides the resources for this task – guidance but not direction. Success, which brings the rewards of recognition, patronage and status, involves conforming to the paradigmatic norms (Mulkay, 1975, p. 515):

> It is clear that the quality or significance of a scientist's work is judged in relation to the existing set of scientific assumptions and expectations. Thus, whereas radical departures from a well defined framework are unlikely to be granted recognition early under normal circumstances, original contributions which confirm to established preconceptions will be quickly rewarded.

Thus the dominant norm of academic life is not one of the five listed above, but conformity. Science is not the constant search for novel discoveries but rather the careful application of agreed procedures to the solution of minor problems in order to extend very slightly on existing well-structured bodies of knowledge. Judgements are being made all of the time, but within an academic environment carefully structured by the training process. Science progresses through sequences of relatively small decisions involved in the solution of relatively minor problems.

The operation of a paradigm is known as *normal science* (Kuhn, 1962, pp. 35–6):

> Perhaps the most striking feature of the normal research problems . . . is how little they aim to produce major novelties, conceptual or phenomenal. Sometimes . . . everything but the most esoteric detail of the result is known in advance . . . the range of anticipated, and thus assimilable, results is always small compared with the range that imaginations can conceive . . . the aim of normal science is not major substantive novelties . . . the results gained in normal research are significant because they add to the scope and precision with which the paradigm can be applied. . . . Though its outcome can be anticipated, often in detail so great that what remains to be known is itself uninteresting, the way to achieve that outcome remains very much in doubt. . . . The man who succeeds proves himself an expert problemsolver.

Within normal science, therefore, the researcher has available:
1 an accepted body of knowledge, ordered and interpreted in a particular way;
2 an indication of the puzzles that remain to be solved; and
3 a set of procedures for puzzle-solving.
Training within a paradigm places the researcher in a fairly deep rut, and provides the tools for extending the paradigmatic body of knowledge (deepening the rut). The result is 'conventional, routinized practice' (Barnes, 1982, p. 11).

Scientists are not omniscient. They do not understand everything – even within a particular paradigm – so that occasionally their predictions prove to be inaccurate. While the process of normal science continues, therefore, slowly accumulating extra knowledge as problems are solved, it sometimes throws up anomalies, findings that are not in accord with the paradigm's assumptions. These must be accounted for (Barnes, 1982, p. 53):

> Puzzle-solving activity frequently attempts to show that what is *prima facie* anomalous is either the spurious product of bad equipment or technique, or a familiar phenomenon in disguise. And most anomalies are successfully assimilated in this way.

Either the work was badly done or the researcher interpreted the results wrongly. Minor adaptation of the paradigm may be needed, but the general process of normal science continues.

Some anomalies cannot readily be accounted for, however, and they continue to worry a few scientists. Their persistence leads to work on alternative paradigms, potential new frameworks that structure knowledge so that there are no anomalies. This is 'extraordinary research', conducted outside the bounds of the accepted paradigm. When it is successful, a 'revolutionary episode' is in progress. Regarding such research, Kuhn (1962, pp. 89–90) notes that:

> Almost always the men who achieve these fundamental inventions of a new paradigm have been either very young or very new to the field whose paradigm they change . . . obviously these are the men who, being little committed by prior practice to the traditional rules of normal science, are particularly likely to see that those rules no longer define a playable game and to conceive another set that can replace them.

The result is an alternative paradigm, which is presented to the adherents of the current normal science. These are asked to make a judgement, not a relatively straightforward one related to their socialization into the accepted paradigm but a major choice between two competing views of their chosen subject. A preference must be expressed. Either the accepted mode of working is to be maintained, despite the anomalies, or a new subculture is to be adopted. The invitation is to discard existing authorities and habits and to take up new ones, which claim to be superior – because they are better predictors of that aspect of the world being studied. If the need for change is accepted, then a revolution in scientific practice occurs; one paradigm is replaced by another.

Kuhn's representation of scientific activity, therefore, is one of 'long periods of dreary conformity interrupted by brief outbreaks of irrational deviance' (Barnes, 1982, p. 13); revolutions are 'irrational' because they involve 'leaps of faith'. During the periods of 'dreary conformity', researchers are trained to employ a proven mode of looking at their subject matter and to use an accepted methodology for solving the problems that they identify. Their problem-solving proceeds in a steady, cumulative manner, adding to the store of knowledge: small modifications may be needed to accommodate minor anomalies that they discover. Very occasionally, anomalies are encountered which cannot be either explained away or accommodated. Some researchers may decide to focus on these, and to develop new paradigms that will account for them, as well as everything else that was already known. When this is achieved, the new paradigm is presented to the research community for approval. A revolution is invited, for the alternative paradigms are incommensurable and only one can be right: the community is asked to accept a new direction to its work. In a sentence, therefore, science proceeds in a steady fashion, along well-trodden paths, with occasional major breaks in its continuity marked by important changes in the organization of its material, in the definition of its problems, and in its techniques for problem-solving.

Criticisms of Kuhn's work

Kuhn's work stimulated many reactions and debates, in part because of its clear suggestion that at least some scientific decision-making was 'irrational'; such a thesis did not accord with the objective neutrality model with which scientists, and philosophers of science, presented themselves. Thus, much of the debate focused on the concept of the revolutionary episode, which is when the 'irrational' judgements must be made. There has also been discussion of the nature of a paradigm, occasioned by Kuhn's relatively cavalier use of the term: Masterman (1970) identifies no less than twenty-one different usages in Kuhn's book, which she groups into three types:

1 the metaphysical, representing a 'world view';
2 the sociological, which provides a set of working habits; and
3 the construct, which is a classic work that provides the tools and procedures for later investigations.

Kuhn himself recognized the existence of these three, and provided two separate definitions which differ in their scale. Thus a paradigm as a *disciplinary matrix* is the sociological definition: 'what the members of a scientific community, and they alone, share' (Kuhn, 1977, p. 460). Secondly, a paradigm may be an *exemplar*, which is a model application to a particular set of problems. In his training, therefore, a researcher is introduced, first, to a world view – that of science. Secondly, the disciplinary matrix, the overarching paradigm, is provided. And finally, the researcher is provided with an 'arsenal of exemplars' (Kuhn, 1977, p. 471) to define the conduct of future work.

Revolutions in disciplinary practice involve a change in its disciplinary matrix. Kuhn's concept of the revolutionary episode has been subject to careful scrutiny (see Stegmuller, 1978), and according to some reviewers it contains several flaws. Watkins (1970), for example, can accept that once a new paradigm has been adopted by a number of the members of a discipline the revolutionary ideas may spread rapidly through the rest of the community. But he cannot accept that a new paradigm is born instantaneously in the mind of the first individual to query the current practice. Instead, he maintains that

> since it takes time – a matter of years rather than hours – to develop a potential new paradigm, heretical thinking must have been going on for a long time before paradigm-change can occur (p. 37).

A revolutionary episode, then, is not a short, sharp debate between the defenders of the old paradigm ('narrow-minded dogmatists') and the protagonists for the new ('religious fanatics'). It is rather a long drawn out process of claim and counter-claim, which ends with one paradigm being accepted as superior to the other.

According to Watkins's view, a discipline can be in a multiparadigm state for considerable periods of time. This is possible in an alternative view of the history of sciences presented by Lakatos (1978). With Kuhn, Lakatos accepts that no theory (or paradigm) is falsified until a better theory is available, but he posits long debates between competing theories. Central to the debates are what he calls *research programmes* (which are similar to Kuhn's paradigms as disciplinary matrices). Each programme has a hard core, an irrefutable central set of beliefs (Lakatos calls it the *modus tollens*). The methodological rules of the programme include a negative heuristic which directs attention away from the hard core. It also includes a positive heuristic, which directs researchers towards the solution of problems in the 'protective belt' that surrounds the core. Anomalies are solved in this protective belt, by the development of auxiliary hypotheses which can account for the observed deviations and so protect the programme's core. Thus progress within a research programme depends on the ingenuity of the scientists in devising hypotheses that are consistent with the core and which can account for anomalies. Success 'hardens' the core, and is measured by the number of 'verifications' or successful predictions, not by failures ('refutations'). The content of the core itself is never in doubt. Research programmes may falter, however. Their positive heuristic may fail to produce successful predictions. When this happens, the programme falls into a degenerative phase, and it is thus ripe for replacement by a new programme which is still in the progressive phase of expanding its content substantially. A programme shift takes place because the new is able to demonstrate its superiority over the old, in much the same way as Kuhn's paradigm shifts, except that the shift usually takes a considerable time. Change is not instantaneous, Lakatos claims (despite its presentation as such in 'scientific folklore': p. 85). Nor are there crucial experiments, the results of which convince a community that one programme is right and the other wrong. (With hindsight, crucial experiments may be identified, but these are not recognized as such at the time when they are reported. There is a danger, Lakatos and others claim, of rewriting history as a series of crucial experiments.) Thus much of the history of a discipline may involve two or more competing research programmes coexisting (Lakatos, 1978, p. 69):

> The history of science has been and should be a history of competing research programmes (or, if you wish, 'paradigms'), but it has not been and must not become a succession of periods of normal science: the sooner competition starts the better for progress.

Lakatos not only recognizes but requires theoretical pluralism (which leads to criticism from Barnes, 1982, that his is a normative view of science, not a positive one like Kuhn's), and within this pluralist situation (Lakatos, 1978, p. 92):

> Criticism of a programme is a long and often frustrating process and one must treat budding programmes leniently.

Although Lakatos denies Kuhn's concept of normal science as one of paradigm dominance within a discipline, he accepts that progressive research programmes comprise relatively routine extensions to knowledge through the testing of new hypotheses derived from the positive heuristic. Refutations are rare; the aim is verification and progress. Popper has argued against this, however, claiming that 'science is essentially critical' (Popper, 1970, p. 55), being characterized not by normal science but by extraordinary research. To him, the normal scientist is really an applied scientist (Popper, 1970, pp. 52–3):

> 'Normal' science . . . is the activity of the non-revolutionary, or more precisely, the not-too-critical professional: of the science student who accepts the ruling dogma of the day; who does not wish to challenge it; and who accepts a new revolutionary theory only if almost everybody else is ready to accept it – if it becomes fashionable by a kind of bandwagon effect. To resist a new fashion needs perhaps as much courage as was needed to bring it about . . . The success of the 'normal' scientist consists, entirely, in showing that the ruling theory can be properly and satisfactorily applied in order to reach a solution of the puzzle in question.

Science, to Popper, consists of bold conjectures and the conduct of experiments designed to refute them. (In his view of science – see p. 72 – hypotheses can never be verified, only falsified.) According to this view, disciplines are in constant revolution as researchers seek to prove that each others' theories are wrong.

Popper's concept of a 'revolution in permanence' has been criticized by Barnes (1982) as normative and not descriptive of science as it is actually practised. Kuhn's (1970, p. 243) response is that:

> [Popper] and his group argue that the scientist should try at all times to be a critic and a proliferator of alternate theories. I urge the desirability of an alternate strategy which reserves such behaviour for special occasions.

According to this alternative strategy, most scientists are 'normal scientists' working routinely rather than critically. This view is supported by Eilon (1975), who has classified management scientists into:
1 Chroniclers, whose role is to describe reality within the constraints of a particular paradigm;
2 Dialecticians, who stimulate debate and progress – a dialectician 'believes it is necessary to debate and argue issues in order to elicit the facts . . . challenging stated views or records, in order to uncover what otherwise may remain hidden from an innocent observer' (Eilon, 1975, p. 361);
3 Puzzle-solvers, who advance the empirical content of the accepted paradigm;

4 Empiricists, who are like chroniclers in their focus on description;

5 Classifiers, who take information generated by empiricists and chroniclers and order it within the paradigm framework;

6 Iconoclasts, who destroy cherished beliefs, the assumptions on which they are built, the deductions, conclusions and interpretations, and expose the incompatibilities between theory and practice, between predicted and observed worlds — such iconoclasts may be destructive only, or they may be constructive in their presentation of alternative theories, in which case they are the agents of paradigm change; and

7 Change-agents, who are the applied scientists concerned not so much with the establishment of theories and facts but rather with the use of existing knowledge to create a better world.

Individual workers may occupy (not necessarily contemporaneously) more than one of these archetypes. If Kuhn and Lakatos are correct, however, most will be chroniclers, puzzle-solvers, empiricists and classifers (Mercer, 1977). The change-agents, too, are conformists, since they seek to apply the accepted normal science, not to challenge it. Popper, then, focuses on a few key scientists, and seeks to characterize the whole of science as if they were typical. (Note, however, that in defending this view, Magee (1975, p. 41) claims that the work of normal scientists can be studied in Popperian terms. They may 'take for granted, in order to solve problems at a lower level, theories which only a few of their colleagues are questioning', but they should be seeking intra-paradigm progress by using the methodology of conjecture and refutation.)

Popper's arguments have been taken much further by Feyerabend (1975). He claims that history is a series of accidents, and that this is how it should be. Similarly the history of science should be allowed to develop as a series of accidents, since this is the best way of ensuring progress. In the past, the main discoveries have been made by individuals who either deliberately or unwittingly flouted the rules but modern scientific education seeks to prevent this by constraining its practitioners into myopic paradigms. Feyerabend claims that the only rule in science should be 'anything goes', in a scientific anarchy, like a political anarchy whose hallmark is (Feyerabend, 1975, p. 187):

> its opposition to the established order of things: to the state; its institutions, the ideologies that support and glorify those institutions.

Like Popper, however, Feyerabend presents a normative model, one that is concerned with major scientific developments and not with their everyday extensions. With regard to these latter, if Lakatos is correct, and not Kuhn — if revolutions take a long time — then this suggests that a discipline need not be characterized by consensus over much of its history. Is it likely that a body of scholars (especially a large body) will agree on ends and means and will not differ at all, except very occasionally and then only for short periods, on fundamental interpretations? From his research on radio astronomy, Mulkay (1978, p. 11) concludes that 'Scientific consensus . . . in a given area of interest is seldom complete'. A discipline may contain several branches (Mulkay, 1975, p. 520):

> In science, new problem areas are regularly created and associated social net-works formed . . . The onset of growth in a new area typically follows the percep-tion, by scientists already at work in one or more existing areas, of unresolved problems, unexpected observations or unusual technical advances, the pursuit of which lies outside their present field. Thus the exploitation of a new area is usually set in motion by a process of scientific migration.

This does not involve a paradigm shift, nor even a degenerating research programme. Rather it indicates the establishment of a new paradigm-exemplar (in Lakatosian terms, a shift to another area of the positive heuristic or protective aureole). A new branch may be identified either by its focus on methods and techniques, or by its attention to, as-yet, relatively ignored subject matter.

The branching process suggested by Mulkay allows for debate and move-ment within a paradigm/research programme without substantial conflict, since the disciplinary matrix/hard core is not being attacked, let alone the 'world view' (see p. 23). Kuhn (1977, p. 462) accepts this, noting that 'Individual scientists, particularly the ablest, will belong to several such [branches] . . ., either simultaneously or in succession', and Suppe (1977b, p. 498) notes that inter-branch consensus is usual:

> to account for normal science all [one] has to assume is that scientists in a particu-lar scientific community are in sufficient agreement on what theory to employ, what counts as good and bad science, what the relevant questions are, what sort of work to take as exemplary, and so forth.

It is the disciplinary matrix/hard core that is central to the paradigm/research programme, not the exemplars and the type of problem that is attacked.

Training in its disciplinary matrix allows researchers to move among branches of a scientific community, therefore. Such branches may even-tually break away, establishing new communities – if not new disciplinary matrices – by what may be termed 'quiet revolutions'. Such breaks are not usually in the interests of the parent community, on political grounds, for they are likely to be against the interests of the disciplinary bureaucracy; separate communities competing for students and research funds, yet cover-ing similar subject-matter, are not favoured. Thus communities may be pre-pared to contain dissenting groups, even potential revolutionaries, rather than risk break-away success. But such dissent is often contained, and may even be repressed. Toulmin (1970, p. 45) notes that the conflict is fre-quently intergenerational:

> perhaps every new generation of scientists having any original ideas or 'slant' of its own finds itself, at certain points and in certain respects, at cross-purposes with the immediately preceding generation.

Its seniors may decide to accommodate the differences, 'for the good of the subject'. But (Mulkay, 1978, p. 116):

> there are now several well documented studies of instances where demonstrably competent scientists have been excluded from a field of study as their ideas have come to diverge from those in the majority.

Thus in some cases the competition between paradigms/research programmes is contained within a discipline. In others, it results in the establishment of separate disciplines, usually after a stage in which the nascent discipline operates as a separate community within the parent body.

Kuhn and the social sciences

These critiques of Kuhn's ideas collectively focus on most of the elements of his thesis. Thus Lakatos questions whether revolutions are achieved in short periods, Popper doubts the existence of normal science, and Mulkay suggests that several paradigms (as exemplars) can coexist within a discipline which is not characterized by consensus. In all cases, there is no questioning the basic world-view, which (implicitly) is that science is the study of an empirical world, in which subject and object can be separated, and that scientific progress is measured by the volume of successful predictions. This suggests that science is an exercise that is independent of the larger society within which the scientists operate. More importantly, as made clear later in this book, it assumes a certain philosophy of knowledge.

This last assumption raises doubts about the validity of Kuhn's thesis – plus those of Lakatos, Popper and Mulkay – to the social sciences. This has not prevented social scientists seeking to fit the Kuhnian model to their disciplines, however, an exercise which Kuhn himself thought to be nonviable and which has drawn scathing criticism (Barnes, 1982, p. 120):

> The popularity of debates about whether sociology has a paradigm, or whether there have been scientific revolutions in economics or in psychology, attests more to the prevalence of intellectual laziness than to the significance of Kuhn's thinking. He himself . . . has stressed that a case for their utility has only been made in the context of the history of the natural sciences.

Kuhn's is a positivist model (p. 71), characterized by its twin emphasis on the dominance of 'super-theories' (disciplinary matrices) and the measurement of success with predictions that form empirical laws. Neither is necessarily relevant to the social sciences. As Gutting (1980, p. 13) points out, the disciplinary matrix element assumes consensus over 'super-theories', which is rare within the social sciences – although some (such as Keynesian economics) have dominated individual disciplines for a time. Replacement of the paradigm concept by that of a research programme can in part counter this, as Blaug (1975) suggests in his outline of Keynesian theory passing through progressive and degenerate stages. But the concept of competing research programmes assumes that the contestants can be assessed on a single criterion (their predictive success). Some social scientific world views do not include predictive ability in their programmes. Thus they are not 'sciences' and cannot be assessed on 'scientific' criteria. Their role in the social sciences must be analysed in a different way.

One potential framework for the analysis of a social science discipline that lacks the particular world-view of Kuhn and his critics has been outlined in Foucault's (1972) *The Archaeology of Knowledge*. This aims to provide a method for the 'pure description' of discourses, which are unified systems of statements (however expressed) that can only be understood within their context (Foucault, 1972, p. 97):

> One cannot say a sentence, one cannot transform it into a statement, unless a collateral space is brought into operation.

Anything that is said or written can only be understood by those privy to the context of the discourse; as with a language, one can only comprehend the words, and the sequence in which they are used, because one understands the rules that govern their use.

According to this view, the history of ideas is the history of discourses, of systems (somewhat akin to languages) used for the discussion of subjects that are defined within the discourses. (There is, for example, no fixed definition of madness; each discourse that discusses it has its own definition and any discussion of madness is particular to that definition and the discourse in which it is set.) The nature of a discourse is deposited in an archive, and it is the function of archaeology to reconstitute the discourse from the archive, to describe what was being done in a particular configuration (Foucault, 1972, pp. 138–40, 157–65). It describes the nature of the discourse without imposing a prescribed framework upon it. As Foucault (1972, pp. 33–4) describes nineteenth-century medical science:

> [it] was characterized not so much by its objects or concepts as by a certain *style*, a certain constant manner of statement . . . medicine no longer consisted of a group of traditions, observations and heterogeneous practices, but of a corpus of knowledge that presupposed the same way of looking at things . . . [in addition it was] a group of hypotheses about life and death, of ethical choices, of therapeutic decisions, of institutional regulations, of teaching models, as a group of descriptions . . . if there is a unity, its principle is not therefore a determined form of statements; is it not rather the group of rules, which, simultaneously or in turn, have made possible purely perceptual descriptions . . .? What one must characterize and individualize is the coexistence of these dispersed and heterogeneous statements; the system that governs their division, the degree to which they depend upon one another, the way in which they interlock or exclude one another, the transformation that they undergo, and the play of their location, arrangement, and replacement. . . .

Discourses, therefore, are sets of mutually agreed rules which govern description and discussion among the members who agree those rules. They are not independent of wider conditions, however. Archaeology may identify separate discursive formations, but these must be related to what

Foucault calls the *episteme*, defined as (Foucault, 1972, p. 19):

> something like a world-view, a slice of history common to all branches of know-
> ledge, which imposes on each one the same norms and postulates, a general stage
> of reason, a certain structure of thought that the men of a particular period can-
> not escape.

In this concept, he appears to suggest that in any period a particular world-
view (macro-paradigm?) predominates and influences (controls?) the contents of individual discourses. Presumably, too, there must be periods of
competition between *epistemes* (which may suggest a similar progress to that
outlined by Lakatos). Foucault does not outline how such change occurs,
however. One is left with the concept of discourses set in a societal matrix,
but with little indication of how the nature of the matrix changes, presumably in part through its interactions with those discourses.

The external environment

The discussion so far in this chapter has suggested that scientific disciplines
(and/or discourses) are communities, small societies that are microcosms of
those within which they are contained. As such, the proper study of how
they operate is sociological, although philosophy may provide a normative
framework for such investigation.

Sociological studies of contemporary societies accept that communities
are not autonomous. Their members are not cut off from the world. (Some of
the earliest, monastic scientific communities were, of course.) They need the
support of the wider society in order to exist, for society employs academics.
And while in part they may be able to impose their own priorities over what
type of work is done, they are strongly influenced by external factors
(Barnes, 1972, pp. 102–3):

> Social, technical and economic determinants routinely affect the rate and direc-
> tion of scientific growth. . . . It is true that much scientific change occurs despite,
> rather than because of, external direction or financial control. . . . Progress in the
> disinterested study [of certain] . . . areas has probably occurred just that bit more
> rapidly because of their relevance to other matters.

Thus the study of a discipline must be set in its societal context. It must not
necessarily be assumed, however, that members of academic communities
fully accept the social context and the directives and impulses that it issues.
They may wish to counter it, and use their academic base as a focus for their
discontent. But the (potential) limits to that discontent are substantial. Most
academic communities are located in universities, many of which are dependent for their existence on public funds disbursed by governments. These
are central bodies in modern societies, which may use their financial power
to influence, if not direct, what is taught and researched on in universities.
And some universities are dependent on private sources of finance, so they
must convince their sponsors that their work is relevant to current societal

concerns. What, then, have those concerns been in the period covered by this book?

World War II is more than a convenient period from which to commence this history of human geography; it marked a major watershed in the development of the societies which are the focus of the book – the United Kingdom and the United States. It cannot be considered in isolation, however. Just as important for the present discussion are the world-wide economic depression which preceded it and the Cold War which followed.

For the first time a major international conflict was not determined solely by the sacrifices of men in battles on land and sea, although there were many such sacrifices during World War II. And the extra dimensions of this war did not just involve the development of air space as a further arena for conflict. The war was fought not only between military forces with guns, but also between scientists, and victory was hastened, if not ensured, by the scientific superiority of the Allied Powers, most obviously at Hiroshima and Nagasaki. It marked the 'coming-of-age' of science and technology. For long these had been major elements in the developing industrialization of the western world, but their dominance was established during the years of conflict, and there was to be no retreat from the many technological advances made by the researchers who assisted in the military effort. Thus the war heralded the era of science and technology, of the dominance of the machine.

Associated with this growth in activity, and also in prestige (with governments and with society at large), was a parallel development of what has become known as social engineering. The major economic depression of the 1930s, finally triggered by the Wall Street crash of 1929, had a massive impact on the nature of government activity and led to the initiation of many measures aimed at the relief of poverty and deprivation and the assuaging of the liberal conscience. In the United States, this was represented by the New Deal legislation of President Roosevelt's governments, which aimed at relief and encouragement to industry and, through the Social Security Act, public support for those who, by no fault of their own, were indigent. Similar measures were introduced by the National Government in Britain; more were foreshadowed by the plans of the war years, such as those promoted in the Beveridge Report on social security, and the landslide victory of the Labour Party in 1945 heralded the introduction of many social democrat policies aimed at giving government a much greater peacetime role in the organization of the economy and society than had ever previously been envisaged.

This development of social engineering was associated with a rising status for the social sciences, and a great expansion in their activities. Economics was the first to achieve prestige, through the contributions of Keynes and others to the solution of the problems of the depression, the organization of the economies during wartime, and participation in the planning of a new world economic order after the war. Others followed. Social psychological research was widely used in the evaluation of personnel by the armed forces and after the war opinion surveys proved valuable to politicians and related groups while market research was increasingly used by industry (along with psychology in its advertising efforts). All of these fields adopted the 'scientific methods' of the more prestigious hard sciences, and their successes were emulated by other disciplines, such as sociology, social administration, and, later, geography. To be scientific was to be respectable and useful.

The war years saw the end of the economic deprivations of the depression, as manufacturing output was boosted to provide the machinery of war. During the cold-war period that followed, military production remained considerable and kept many people in work. Further, there were many years of doing-without to be compensated for, and with full employment, government direction of, and increasing involvement in, economic affairs, and the need to re-equip industries, the two decades following the war were characterized by an economic boom in the western world. Apart from the greater government involvement, this era in industrial development was marked by another major new characteristic, the development of the giant firm, including the multi-national. The concentration and centralization of capital proceeded rapidly, probably more rapidly than during any preceding period; the average size of firms and factories increased and the economy of the world became dominated by a relatively small number of concerns, which were usually encouraged in their activities by governments.

Rebuilding the ravaged war arenas of Europe also placed new demands on societies, and the planning profession emerged from earlier obscurity to take on a major role in preparing the blueprint for a new social order. The need for such action had been realized during the war years in the United Kingdom with the preparation of a series of reports concerned with future land-use patterns, and with the spatial distribution of economic activity, at local and regional scales. Cities were to be rebuilt; New Towns were to be constructed; a more balanced inter-regional distribution of industry and employment was to be ensured; agricultural land was to be protected; and residential environments were to be improved: all of this made great demands on social scientists, as well as engineers. The greater degree of commitment to the private ownership of land in the United States led to a slightly slower acceptance of the need for spatial planning there, but its heyday came with the rapid growth of problems involved in catering for the upsurge of ownership and use of the automobile: transport planning and engineering soon became growth industries, allied with the automobile industry and the companies which constructed the major highways. Again, this became a major sphere of government activity.

For all of these tasks – economic growth and planning; spatial planning; social administration; technological change; management, etc. – there was a need for educated personnel, and the universities received unprecedented demands for their graduates to serve the new needs of society. Education expanded rapidly. The existing universities and colleges grew and many more were founded. Science and social science departments expanded to meet the need for more students. The extra staff were involved in research, which increased the tempo of paradigm development and questioning, and the products of the research were wanted by governments and businesses to assist in the achievements of their aims. Rather than the places where a small elite were educated and a few privileged individuals followed their research interests, the universities became centres of society's development – the 'white-hot technological revolution' which Harold Wilson promised the British in the early 1960s. Research projects became bigger, supported by large grants from outside bodies and carried out by specially employed graduates, and the rate and volume of publication increased exponentially.

The years from 1945 to about 1965 can be characterized as a period of scientific and technological dominance, therefore. The problems of production had been solved, it was claimed, in that enough goods and services could be provided to satisfy all. The problems of distribution were still being solved, for as yet there was still inequality of provision at all spatial scales. But these could be handled, it was argued, and the prospect of a prosperous and healthy life for all was widely canvassed. Academic disciplines were contributing substantially to this problem-solving, by their own scientific progress. Advances in the natural sciences and technologies were solving the problems of production – of food, housing, and consumer goods – as well as of ill-health. And advances in the social sciences were aiding in the management of success. Investment in education was thus investment in social progress (as well as an investment in the life chances of the individuals involved).

Despite the successes of this initial post-war period – best characterized, probably, by Harold Macmillan's 1959 election slogan 'you've never had it so good' – doubts persisted. By the mid-1960s they were growing, and by the early 1970s they were having a major impact on the world scene. The seeds of the doubt were many. Initially they were focused, especially in the United States, on the problems of nuclear weapon development and of war, particulary the increasingly unpopular conflict in Vietnam and surrounding countries where technology was not carrying all before it against strong popular resistance and the casualty rates were more and more deplored. Beneath the humanitarian concerns were doubts about the inequalities that continued, both at a world scale and within the 'successful' capitalist societies. Poverty was not being alleviated; if anything the disparities between rich and poor were being extended, and absolute standards of living remained appallingly low in many parts of the world, condemning the majority of the population to short lives of continuous deprivation. The prospects for solving these problems were much less rosy than they had been a few years earlier.

Within the successful societies, it was increasingly realized that success was being bought at considerable cost. Scientific and technological advancement required the dominance of the machine and the large factory. Work for large proportions of the labour force was being made more repetitive and boring, as skilled tasks were taken over by automated production and assembly lines. Alienation of the individual from society was increased. And within society, particular groups suffered more than others, as the results of prejudice and discrimination. Ethnic minority groups were the main sufferers, along with women in most societies, and they were the focus of civil rights movements. Finally, interest was kindled in the growing despoliation of the environment to fuel the production goals of advanced capitalist societies.

The problems of the dignity of the individual, the repression of minorities, the quality of life and the depletion of environmental resources were not new in the late 1960s. Nor were they first realized then. What was realized was that the form of social 'progress' advanced during the previous decades was exacerbating and not solving such problems. As the realization grew, so the proposed solutions varied. To some, the problems could be solved by greater state involvement, at an international and national scale. Human and civil rights must be protected; greater equality must be achieved through the redistribution of wealth; the environment must be conserved, and where necessary preserved. In the language of the previous section, the research programme is maintained, but major efforts are made within its positive heuristic to solve the many anomalies. To others, this solution was insufficient. It would only lead to new anomalies because, according to the developing critique of capitalist systems, these are necessary to social 'progress': capitalism, it was argued, only survives on inequalities, on alienation, and on the rape of the environment. Only a new research programme could change society for the better.

Within society, the educational institutions were the centres for much of this developing concern. There were major confrontations between students and other elements of society in the late 1960s, for example, in particular at Berkeley, Chicago, London and Paris. The student body was in the forefront of the anti-war movements. In addition to these explicit statements of concern, there was academic consideration of the problems, notably in the social sciences which were concerned with the 'managerial' issues, which were much more crucial than the 'production' issues covered by the natural sciences and technologies. A threefold division developed. In one, the need for social scientists to become more active in developing solutions to the problems of distribution and environmental depletion was advanced: social science must become more 'relevant', more 'policy-oriented', within the constraints of the current societal 'research programme'. A second argued for greater concentration on the problem of the alienation of the individual, who should be released from the overbearing dominance of the machine and the big organization and should be encouraged to take a much greater part in creating her/his own life. The individual was to be protected against the increasingly distrusted expert. Finally, a third group developed a critique of capitalist society, seeking to show that while specific problems may be soluble this would merely lead to others, while the general problems would remain because they are endemic to that mode of social organization. The call was for major social reform – to some, revolution – as the only long-term solution to the problems of human dignity and inequality.

The force of these arguments can be identified in a variety of ways, not least the declining popularity of scientific and technological subjects among students and the growing demand for places in the social sciences and humanities. Soon, however, a further major problem arose, because the capitalist world spent much of the 1970s in economic recession. To some, this was brought about by the decision of the Arab countries during the 1973 Yom Kippur war to use oil as an economic weapon. The price of this vital raw material was increased many-fold in a few years, with major impacts on all economies. But to others, the recession was already set in – as illustrated by Britain's problems in the late 1960s – and the Arab decision was just a major additional stimulus.

To some social theorists, recession and its major impacts (notably unemployment) should accelerate the demands for reform and revolution. But there is little evidence of this in the countries studied in this book: indeed, at the end of the 1970s both the United Kingdom and the United States elected right-wing governments dedicated to economic regeneration by a reduction in public expenditure and a liberalization of those capitalist forces that produce the inequalities so widely condemned only a decade earlier. Education has been subject to significant policy changes throughout the 1970s because of this shift (which was prefaced by earlier governments, and only extended by those currently in power). For the first time for more than two decades, expenditure on higher education was to be cut, and the numbers of students taking undergraduate degrees and post-graduate training were reduced. The cuts were selective with relative protection for science and technology – growth in which was seen as necessary for economic progress – and substantial reductions in the social sciences, believed by many to be the homes of left-wing radicals and the fomenters of discontent. Research funds were similarly cut and redistributed. The reaction to these policies has been in part an attempt to defend academic freedom and independence and the need for a 'critical conscience' within society. But it has much more been a desire to reorient work within disciplines to make it more 'relevant' to current societal concerns.

The educational system within which the research components of academic disciplines are located has been closely involved in economic, social and political change during the last forty years. It benefited from the boom years of the first two decades. Expansion was rapid and academic activity was considerable. In the later years, it has been cut back, as – to some decision-makers at least – expansion was seen to have been a luxury that could be ill afforded. Economic progress did not require large numbers of students and potential researchers being trained in disciplines with little relevance to perceived societal needs and working on topics that were critical of societal structure. Education was to be cut back, and investment programmes more closely geared to those perceived needs. Disciplines and scholars were to prove their relevance and to sell their skills in the marketplace. Academic freedom was not entirely invaded, but was to be curtailed, simply by denying it resources if it was felt that freedom was being abused. The result is that after participating gleefully in the booms of the 1950s and 1960s, academia is now in deep depression, suffering internal and external crises of confidence and subject to considerable political direction.

Conclusions

The thesis of this chapter is that the history of an academic discipline must be set in a context comprising three elements: the occupational structure; the organizational framework for research; and the societal environment. These three interact in a variety of ways. The occupational structure is very much constrained by the societal environment, for example, as indicated by the negligible promotional opportunities in British universities in the later 1970s/early 1980s because of the cuts in educational funding. Similarly, the framework for research, although established by and for academics, is subject to societal support. Some frameworks are much more acceptable than others, and so are much more likely to receive the needed public finance.

In the following chapters, the content of human geography in Britain and North America since World War II is reviewed, within the context set by the discussion here. The emphasis is on 'extraordinary research' rather than the cumulative achievements of 'normal science', stressing the debates over the need for change in how research is done in human geography. Thus the term paradigm is freely used, while not signifying adherence to Kuhn's thesis (see Chapter 7). No attempt is made in the next five chapters to test the models outlined in the section above on 'The academic working environment', although the ideas outlined there have clearly influenced the organization of the book. The main purpose is to present a reasoned chronology of debates in human geography (the discipline being defined as comprising that which is claimed as human geography), based entirely on the published record. The relationship between this chronology and the models of scientific progress only resurfaces in the final chapter, where an attempt is made to evaluate and structure the chronology.

2

Foundations

Although this book is about human geography since 1945, the discussion of that period must be preceded by a brief outline of the nature of the discipline in the previous decades. Such a foundation is needed for a variety of reasons. The first is that although 1945 was something of a watershed year in many aspects of the social, economic and intellectual life of the countries being considered here, it did not mark a major divide in the views on geographical philosophy and methodology. Not surprisingly, the war years were not a major period of intra-discipline academic debate. Most academics spent much of the war either on active service or in associated intelligence activities (some of those involved in the latter retained their teaching commitments); the everyday activities of teaching, pure research, and administration were replaced by commitment to the war effort. It took a few years for academic life to return to something like normality, to assimilate the large numbers of new staff needed to replace the losses of the war years and to teach the backlog of students, and to react to the new social and economic environments.

A second reason for surveying the period preceding that being studied relates to the processes of change in academic work. New paradigms are created as reactions to those currently in favour, and not as inventions in an intellectual vacuum. Thus the post-Second World War changes were reactions to the philosophies and methodologies which had been developed and taught in the preceding decades; the nature of the reactions cannot be studied without some knowledge of what went before.

Finally, change is not instantaneous in academic life. A new research programme usually takes years to mature, while experimentation with alternatives takes place, the programmatic statements are written, and the converts are won over by the prophets of the new approach. Meanwhile the current paradigm continues (or paradigms, if there are several with considerable support). Its adherents continue to work in their accepted ways, conducting research, publishing, and teaching generations of undergraduates according to the conventional wisdom. Even when a new paradigm has been crystallized, it may be that it must coexist for several years with its predecessor, as a competitor for the support of academics and students. This may be especially characteristic of the social sciences, in which interpretations of data are frequently more subjective than is the case with the physical sciences; it is quite feasible for two or more separate world views to find adherents at the same time, even in the same academic department.

Geography in the modern period

The hallmark of an academic discipline, according to one of geography's chroniclers (James, 1972), is that it has an educational organization which provides a specialist training in the subject. James dates the beginning of such an organization for geography at around 1874, when the first university geography departments were established in Germany: Britain and the United States followed a little later, with the main developments coming in the twentieth century. Before 1874, geography was a subject investigated either by amateurs or by scientists trained in other fields. With its own specialized institutional training, geography left its classical age and entered what James terms its modern period, which lasted for about eighty years, being superseded after 1945 by what James calls its contemporary period.

James's modern period is virtually co-extensive with the decades surveyed by Freeman (1961) in his *A Hundred Years of Geography* which, with James's book, is one of the few relatively recent attempts to provide a history of the discipline (see also Freeman, 1980a). Freeman identified six main trends in the geographical literature.

1   *The encyclopaedic trend,* associated with the collection of new information about the world, particularly areas little known to the residents of western Europe and North America. Although the great age of discovery was over, and by the late nineteenth century much of the world had been visited by Anglo-Saxon explorers, there were still vast tracts, notably in Africa, which if not *terra incognita* were extremely empty on contemporary maps. Indeed, at the beginning of geography's modern period much of the North American continent itself remained to be settled by permanent farms.

2   *The educational trend,* which, as James stresses, characterizes an academic discipline needing to propagate its knowledge, establish its relevance, and ensure its reproduction. Much work was undertaken to achieve a solid foundation of geographical work in schools, colleges and universities, involving both proselytizers and the architects of curricula (Freeman, 1980a, 1980b).

3   *The colonial trend* reflects a major environmental preoccupation during the early decades of the modern period, especially in Britain whose empire was being consolidated and developed into a spatial division of labour based on its metropolitan hub and covering a considerable proportion of the earth's surface. Organization of the commercial world required a great deal of information about the various countries concerned, the provision of which became a major task of geographical research whilst its propagation was the keystone of geographical education.

4   *The generalizing trend* describes the use to which data collected in the

encyclopaedic and colonial traditions were increasingly employed. Academic study involved more than the collection and collation of facts: these had ·to be interpreted, and the methods and aims of such interpretation defined the early paradigms of the discipline's development.

5 *The political trend* was reflected in the contemporary uses made of geographical expertise. Isaiah Bowman was a chief adviser to Woodrow Wilson at the conferences which re-drew the map of the world after the First World War, for example, and the work of geopoliticians such as Haushofer was influential on the *lebensraum* ideology of Nazi Germany (James and Martin, 1981).

6 *The specialization trend* was a reaction to the growth of knowledge and the inability of any one individual to master it all, even within the single discipline of geography. Prior to the modern period, it was possible for scientists and other academics to be extremely catholic in their interests and expertise, but as the volume of research literature increased and the techniques of investigation demanded longer and more rigorous training so it became necessary for the individual to specialize, first as a geographer, and then within geography, focusing either on one substantive area or on a particular region of the earth's surface.

Some of these trends represent philosophies, some methodologies, and some ideologies with regard to the purpose of academic geography. From them it is possible to identify three paradigms which characterized the modern period, both of human geography as a whole and of its component parts, such as urban geography (Herbert and Johnston, 1978). Discussion of these three occupies the remainder of this chapter.

Exploration

The first of the paradigms was carried over into the modern period from the classical, for exploration was the major activity recognized as geography through most of the nineteenth century. The collection and classification of information about 'unknown' parts of the earth (unknown, that is, to western Europeans and North Americans) was undertaken by geographers. Many of their expeditions were financed through the geographical societies which were founded during that century (Freeman, 1961, 1980b); these, in turn, obtained money from commercial as well as philanthropical sources, for the information gathered was of great value to the mercantile world. As well as supporting and sponsoring exploratory expeditions, the geographical societies also undertook major educational roles. Their lecture meetings provided opportunities for the general public to see and hear of the new discoveries, and their officers worked hard to establish the teaching of geography in schools and universities: the Royal Geographical Society of London (RGS), for example, was involved in discussions which led to the establishment of geography teaching at England's two oldest universities, Oxford and Cambridge (Stoddart, 1975a; Freeman, 1980b; Cameron, 1980).

The importance of exploration declined as geography matured in its new academic-discipline status during the early twentieth century, although in 1899 Halford Mackinder felt it necessary to establish his credentials as a geographer by becoming the first person to climb Mt Kenya. There was still much *terra incognita*, however, and the geographical societies maintained their interest in and sponsorship of expeditions. Indeed, the RGS still acts as a major sponsor of scientific expeditions large and small and its major publication – the *Geographical Journal* – reports the findings of many of them. The nature of the work undertaken is very different from that of a century ago in most cases, reflecting developments in scientific technology and the available store of knowledge, but basic activities such as accurate map-making are still crucial to many successful expeditions. Many of the Society's meetings still present, in words and pictures, the results of expeditions and travels to all parts of the world, meetings which remain very popular with its large membership that is dominated by non-academics.

The American Geographical Society in New York is the American counterpart of the RGS. It has probably retreated a little further from the exploration role which it too focused on early in the modern period, and it now sponsors research in many other areas of geography, while still providing major support for investigations of relatively unstudied parts of the globe, such as the Arctic. The exploration tradition is maintained in the United States by the National Geographical Society and its popular journal, the *National Geographical Magazine*. Other societies have been established to take over some of the other professional roles: there are separate academic bodies in both the United States and Great Britain (the Association of American Geographers and the Institute of British Geographers) as well as those which concentrate on geographical education (the National Council for Geographical Education and the Geographical Association).

Although most of it was not strictly exploration, the work summarized by Freeman under the colonial trend can also be classified within this paradigm, since its aims were the collection, collation and dissemination of information. Much of the material put together was about commercial activities and infrastructure, as in volumes such as Chisholm's *Handbook of Commercial Geography* (first edition, 1899) and *Gazetteer of the World* (1895), which were aimed at the world of commerce, with companion volumes for schools (Wise, 1975). Their content comprised statistics and descriptions of production and trade, and a training in this type of geography was boring to many with its focus on the assimilation of large bodies of factual knowledge ('capes and bays' geography). But the existence of this geographical expertise was widely recognized, and was called on at the end of the modern period when geographers were made responsible for the preparation of intelligence reports about areas in which allied troops were likely to be engaged, work characterized by the set of British Admiralty Handbooks.

Environmental determinism and possibilism

These competing paradigms respresent the first attempts at generalization by geographers of the modern period. Instead of merely presenting information in an organized manner, either topically or by area, geographers began to seek explanations for the patterns of human occupation of the earth's surface. The major initial source of their explanations was the physical environment, and a theoretical position was established around the belief that the nature of human activity was controlled by the parameters of the physical world within which it was set.

The origins of this environmental determinism lie in the work of Charles Darwin, whose seminal book *On the Origin of Species* (first published in 1859) influenced many scientists. His notions regarding evolution were taken up by the American geographer William Morris Davis in his famous cycle-of-erosion model of landform development. Ideas of natural selection and adaptation formed the basis of statements regarding environmental determinism, including one by Davis (1906) whose programmatic paper identified the core of geography as the relationship between the physical environment, as the control, and human behaviour, the response (Stoddart, 1966; Martin, 1981).

Chief among the early nineteenth-century environmental determinists was the German geographer Ratzel, whose American disciple Ellen Churchill Semple opened her book *Influences of Geographic Environment* (1911) with the statement that 'Man is the product of the earth's surface'. In some hands, the environmental influences adduced were gross, and with hindsight it is hard to believe that they could have been written and taken seriously; a brief survey by Tatham (1953), for example, illustrates the extent to which authors were prepared to credit all aspects of human behaviour with an environmental cause.

Reaction to the extreme generalizations of the environmental determinists led to the development of a counter-thesis, that of possibilism, in which man was presented as an active rather than a passive agent. Led by French geographers, followers of the historian Lucien Febvre, possibilists presented a model of man perceiving the range of alternative uses to which he could put an environment and selecting that which best fitted his cultural dispositions. Taken to extremes, this paradigm could be as ludicrous as that which it opposed, but in general the possibilists recognized the limits to action which environments set, and avoided the great generalizations which characterized their antagonists.

Debate over environmental determinism and possibilism continued into the 1960s (Lewthwaite, 1966) and was actively pursued in the United Kingdom in the first decade after the Second World War. (Spate, 1957, for example, proposed a middle ground with the concept of 'probabilism'.) The determinism cause was continued in the period between the world wars by writers such as Ellsworth Huntington, who advanced theories relating the course of civilization to climate and climatic change. Perhaps the most doughty of all advocates was the Australian Griffith Taylor, whose determinist views so angered politicians interested in the settlement of outback Australia that he was virtually hounded out of his homeland (Powell, 1980a). Taylor's reaction to Tatham's (1953, p. 150) statement that when industrialists decide where to locate a new factory 'Geographical controls are rarely mentioned' was, 'Surely this definitely illustrates the stupidity of the owner!' His case was that the possibilists had developed their arguments in temperate environments such as that of northwestern Europe, which do indeed offer several viable alternative forms of human occupance. But such environments are rare: in most of the world − as in Australia − the environment is much more extreme and its control over man that much greater accordingly He coined the phrase 'stop-and-go' determinism to describe his views. In the short term, man might attempt whatever he wished with regard to his environment, but in the long term, nature's plan would ensure that the environment won the battle and forced a compromise out of its human occupants.

Many debates begin as two opposing, extreme positions, and end as a compromise accepted by all but the most fervent devotees of either polar position. Thus the lengthy discussion among geographers about whether man is a free agent in his use of the earth or whether there is a 'nature's plan' slowly dissolved as the antagonists realized the existence of merits in each case. (And some geographers proceeded independently to study men-environment interactions outside the confines of these debates: see Fleure, 1919.) But while environmental determinism was a view strongly held and widely preached by geographers about whether man is a free agent in his use of the earth, respect for the discipline declined somewhat in the eyes of the academic community at large, who rejected the paradigm. As a consequence, geography's next paradigm, which had strong roots in environmental determinism, was very much an introspective and conservative one.

The region and regionalism

This, the third of the paradigms, dominated British and American geography for much of the first half of the present century. Like environmental determinism, it too was an attempt at generalization, but at generalization without structured explanation, and thus of a very different type from the increasingly discredited law-making attempts of the previous paradigm. Much of the early development took place in Britain, and involved work at two scales (Freeman, 1961, p. 84). At the large scale were efforts, such as Herbertson's (1905), to divide the earth into major natural regions, usually on the basis of climatic parameters and thus having some links with the earlier determinism. At the smaller scale, the aim was to identify individual areas with particular characters:

The fundamental idea was that the small area would legitimately be expected to show sime distinct individuality, if not necessarily entire homogeneity, through a study of *all* its geographical features — structure, climate, soils, vegetation, agriculture, mineral and industrial resources, communications, settlement and distribution of population. All these, it has often been said, are united in the visible landscape, linked into one whole and dependent one on another. And more, every area, save those few never occupied by man, has been influenced, developed and altered by human activity, and therefore the landscape is an end-product, moulded to its present aspect by successive generations of people. The practice has therefore been to take an evolutionary view and . . . to attempt to reconstruct the landscape as it was a hundred, or a thousand years ago (Freeman, 1961, p. 85).

Some of this work, exemplified by Herbertson's (1905), was the precursor of the ecosystem concept.

Hartshorne and the American view

The ideas and methods of regional geography were taken up a little later in the United States. In the late 1930s, however, two non-geographers published a major survey of American regionalism (Odum and Moore, 1938) and in 1939 the Association of American Geographers published a monograph — Richard Hartshorne's *The Nature of Geography: A Critical Survey of Current Thought in the Light of the Past* — which rapidly established itself as the definitive statement of the paradigm. As Hartshorne (1948, 1979) later made clear, there was much debate among American geographers during the 1930s (most of it apparently unpublished) about the nature of their discipline. Hartshorne was concerned about both tone and content of the debate, and in 1938 submitted a paper to the *Annals*, as a contribution to the philosophical discussions. He then proceeded to Europe for fieldwork on boundary problems, as part of his ongoing research into political geography. This work was frustrated by the political situation, and so he spent his time reading European, mainly German, work on the nature of geography. He used this to extend his 1938 paper, adding the sub-title; the result was a 'paper' of 491 pages (some 230,000 words) which became the major philosophical and methodological contribution to the literature of geography in English then available.

A synopsis of Hartshorne's scholarship, and his interpretations of the scholarship of others, notably Hettner, is not possible in a few paragraphs, and only the main conclusions can be stressed here. Hartshorne argued forcefully that the focus of geography is areal differentiation, the mosaic of separate landscapes on th earth's surface. Thus the discipline is:

a science that interprets the realities of areal differentiation of the world as they are found, not only in terms of the differences in certain things from place to place, but also in terms of the total combination of phenomena in each place, different from those at every other place (p. 462) so that

> geography is concerned to provide accurate, orderly and rational description and interpretation of the variable character of the earth surface (p. 21)

and it

> seeks to acquire a complete knowledge of the areal differentiation of the world, and therefore discriminates among the phenomena that vary in different parts of the world only in terms of their geographic significance – i.e. their relation to the total differentiation of areas. Phenomena significant to areal differentiation have areal expression – not necessarily in terms of physical extent over the ground, but as a characteristic of an area of more or less definite extent (p. 463).

According to this view, the principal purpose of geographical scholarship is synthesis, an integration of relevant characteristics to provide a total description of a place – a region – which is identifiable by its peculiar combination of those characteristics. There is then, according to Hartshorne, a close analogy between geography and history; the latter provides a synthesis for 'temporal sections of reality' whereas the former performs a similar task for 'spatial sections of the earth's surface' (p. 460).

Hartshorne also indicated the methodology to be used for this integrating science aimed at orderly description of the earth's surface. To him, 'the ultimate purpose of geography, the study of areal differentiation of the world, is most clearly expressed in regional geography' and so accepted procedures were necessary for regional identification. Regions are characterized by their homogeneity on prescribed characteristics, selected for their salience in highlighting areal differences. Two types of region were identified; the *formal region* (or uniform region) in which the whole of the area is homogeneous with regard to the phenomenon or phenomena under review, and the nodal or *functional region* in which the unity is imparted by organization around a common node, which may be the core area of a state or a town at the centre of a trade area. Identification of such regions

> depends first and fundamentally on the comparison of maps depicting the areal expression of individual phenomena, or of interrelated phenomena . . . geography is represented in the world of knowledge primarily by its technique of map use (pp. 462–4).

Hartshorne placed his emphasis on map *use*. Although it is valuable for geographers to know something about the preparation and construction of maps, the sciences of surveying and map projections are of only secondary interest to them; the prime task of the geographer is in the interpretation of maps, and increasingly, from about 1940 on, of various forms of aerial photograph. Much of this information to be interpreted may have been placed on the maps by geographers during their fieldwork, and the role and nature of fieldwork were of considerable interest to American geographers during the period when Hartshorne was developing his ideas.

Preparation of the materials for a regional synthesis required collection both from other sciences specializing in certain phenomena (though usually not their areal patterning) and from the topical systematic specialisms which complemented, but which were eventually subsidiary to, regional geography. Physical, economic, historical and political were the main systematic subdivisions recognized within geography at the time Hartshorne wrote, although a later survey, set firmly within the regional paradigm, identified many other 'adjectival geographies', including population, settlement, urban, resources, marketing, recreation, agricultural, mineral production, manufacturing, transportation, soils, plant, animal, medical, and military, plus climatology and geomorphology (James and Jones, 1954). A number of these were of only minor importance, however, so that despite the apparent diversity of interests among geographers of the time, the 'classic' regional study usually followed a sequence comprising physical features, climate, vegetation, agriculture, industries, population and the like (Freeman, 1961, p. 142) and summarized by a synthesis of the individual maps to produce a set of formal regions.

To most geographers of the period spanning World War II, and notably those who contributed to the survey edited by James and Jones (1945), regional geography was at the forefront of their discipline's scholarship and systematic studies were the providers of information for that enterprise (it still is to some; Hart, 1982): thus to James, 'Regional geography in the traditional sense seeks to bring together in an areal setting various matters which are treated separately in topical geography' (1954, p. 9). Urban geographers studied towns because they 'constitute distinctive areas' (Mayer, 1954, p. 143), in line with the regional concept; political geographers studied the functions and structures of an area 'as a region homogeneous in political organization, heterogeneous in other respects' (Hartshorne, 1954, p. 174); and in defining the 'new' field of social geography, Watson (1953, p. 482) saw it 'as the identification of different regions of the earth's surface according to associations of social phenomena related to the total environment'. Each of these topical specialisms, therefore, sought to produce its own regionalization (notable in this was the work of agricultural geographers, especially O. E. Baker, in a series of papers published in *Economic Geography* during the 1920s and 1930s outlining the agricultural regions of various parts of the world). It had its links with the relevant systematic sciences – social geography with sociology, for example – and the key differentiating factor between the two was the geographer's focus on the region, the single-attribute region of his specialism and the multi-attribute region in the synthesis of his work with that of others to produce regional geographies.

Given this focus on the region, it is not surprising that the literature of the paradigm contained many contributions discussing the nature and delimitation of such homogeneous areas, for virtually every region was in effect a generalization, complete homogeneity being very rare over more than a small area. As already indicated, British geographers were active early in the definition of large-scale regions, usually based on climatic parameters. Much effort was made to develop methods to define multi-attribute regions; in agricultural geography, for example, it culminated in the statistical method developed by Weaver (1954). But at the small scale it was widely accepted that regional delimitation should be based on personal interpretation of landscape assemblages. For this, the model was the work by the French geographer Paul Vidal de la Blache and his followers on the *pays* of their homeland, small regional units with distinct physical characteristics, notably in soils and drainage, and associated agricultural specialisms (Buttimer, 1971, 1978a).

One systematic specialism which stood slightly apart from the others was historical geography, the study of which was based on the argument that investigations of genesis were needed in order to comprehend the regional patterns of the present. Two approaches to historical geography can be recognized in its literature from the 1920s on. The first, often thought of as the British approach and closely associated with the work of H. C. Darby, involved the detailed study of past geographies (Perry, 1969). This was done in a series of cross-sections, whose locations in time were almost always determined by the available source material, such as the Domesday Book of *c*.1086 which was analysed in great depth by Darby and his associates (culminating in Darby, 1977; Perry, 1979). These cross-sectional analyses, complete with their regionalizations in many cases, were linked together by a narrative outlining the changes between the periods studied: most emphasis was placed on the cross-sections, however, for which data allowed analysis rather than interpretation (see Darby, 1973).

The second approach was largely American in its provenance, and centred upon the works of C. O. Sauer and his associates. The focus was the study of ongoing processes leading to landscape change up to, and including, the present and beginning at the prehuman stage of occupance (Mikesell, 1969): most of the work was conducted either outside the United States itself (particularly in Latin America) or in the less industrialized parts of that country. Sauer's (1925) first methodological statement constrained geographical endeavour closely to the generic study of landscapes, with emphasis on their cultural features (although in fact work was also done on the borderlands between geography and botany); there was no glorification of the region, however. In his later sermons – as he called his methodological and philosophical statements – Sauer (1941, 1956) encouraged research over a much wider field, but emphasized the study of cultural landscapes and the links which he had forged with anthropology, to produce a creative art-form whose hallmark was that it was not prescribed by pattern or method: the human geographer is obliged 'to make cultural processes the base of his thinking and observation' (Sauer, 1941, p. 24). The work, as undertaken by Sauer and his students, involved neither detailed reconstruction of past geographies nor close consideration of regional boundaries: instead it led to a catholic historical geography whose rationale (Clark, 1954, p. 95) was that:

> through its study we may be able to find more complete and better answers to the problems of interpretation of the world both as it is now and as it has been at different times in the past.

Not all American historical geographers followed this lead – Brown (1943), for example, worked on detailed reconstructions of past periods – but the 'Berkeley School' which Sauer founded and led for almost five decades had many followers and a point of view within the prevailing academic orientation of geography which almost made it a separate paradigm within a paradigm, focused on a single iconoclast (Hooson, 1981). Sauer's influence was carried on by his students, notably Leighly, Parsons, and Clark (Bushong, 1981).

The British view

British geographers seem to have been less concerned with philosophical and methodological debate than their American counterparts during the 1920s, 1930s, and 1940s (though see the exchange in the *Scottish Geographical Magazine* during the late 1930s, initiated by Crowe, 1938). They seem to have been more pragmatic in their work, less prone to contemplate the nature of their subject and more prepared, perhaps, to adopt the well used adage that 'Geography is what geographers do'. But they too accepted that the *raison d'être* of geography was synthesis, the integration of the findings of various systematic studies with a strong emphasis on genesis, as in the studies of geomorphology and historical geography (Darby, 1953). According to Wooldridge and East (1958):

> geography . . . fuses the results, if not the methods, of a host of other subjects . . .
> [it] is not a science but merely an aggregate of sciences (p. 14) its *raison d'être* and intellectual attraction arise in large part from the shortcomings of the uncoordinated intellectual world bequeathed us by the specialists (pp. 25–6)
> in its simplest essence the geographical problem is how and why does one part of the earth's surface differ from another (p. 28)

All of these statements indicate a strong trans-Atlantic common body of opinion, although, despite a statement that 'The purpose of regional geography is simply the better understanding of a complex whole by the study of its constituent parts' (p. 159), the British writers did not elevate the regional doctrine as much as did their American counterparts. Nevertheless, Wooldridge (1956, p. 53) wrote in 1951 that

> the aim of regional geography . . . is to gather up the disparate strands of the systematic studies, the geographical aspects of other disciplines, into a coherent and focused unity, to see nature and nurture, physique and personality as closely related and interdependent elements in specific regions

and that in any department of geography each member of the staff should commit himself to the study of a major region (p. 64).

One major difference between British and American geography by the 1950s was in their attitudes to physical geography, the study of the land surface, the atmosphere and the oceans, and their faunal and floral inhabitants. Both countries had strong traditions of work on these topics, and many geographers had academic roots in the associated field of geology. But in North America (the United States much more than Canada) this tradition had slowly dissolved and interest in the physical environment, and particularly its understanding as against its description, waned. This may have been a consequence of the excesses of environmental determinism, and a subsequent desire to remove all traces of that connection and to see man as the formative agent of landscape patterns and change: associated with this was probably the attempt to redefine geography in the 1920s as the study of human ecology, in which man is seen as reacting and adjusting to environments while at the same time attempting to adjust the environment to his own needs (Barrows, 1923). Thus with regard to geomorphology – the science of landform genesis – Peltier (1954, p. 375) wrote:

> the geographer needs precise, factual information about particular places. What landforms actually exist in a given area? How do they differ? Where are they? What are their distribution patterns? The geomorphologist may concern himself with questions of structure, process, and stage, but the geographer wants specific answers to the questions: what? where? and how much?

What geographers were interested in, according to this view, was the geography of landforms: geomorphology, the genetic study of landforms, was a part of geology and, unlike historical geography, was deemed irrelevant to the geographical enterprise. Similar reactions saw the wholesale removal of climatology and biogeography from American geographical curricula, and their replacement by introductory courses in physical geography which described landforms, climates, and plant assemblages – usually in a regional context – but paid little or no attention to their origins. (on recent trends, see Marcus, 1979.)

This American trend was not repeated in Britain where, according to Wooldridge and East (1958, p. 47):

> To treat geography too literally as an affair of the 'quasistatic present' is to make both it and its students seem foolish and superficial. It is true that our primary aim is to describe the present landscape; but it is also to interpret it.
> . . . Our study has therefore always to be evolutionary. . . . It is unscholarly to take either landforms or human societies as 'given' and static facts, though we must not let temporal sequences obscure spatial patterns.

Thus, certainly in the 1950s, geography students at British universities rarely specialized, except perhaps in the final year of their course, in either physical or human geography. Both were considered essential parts of a geographical education, as contributions to the genetic study of regional landscapes which was the integrating focus of geographical scholarship. As researchers, most British geographers specialized in either physical or human geography (though rarely exclusively so), but almost all had a regional specialism as well, in which they 'integrated' studies from 'both sides' of their subject, as widely illustrated in the regional textbooks of the period.

Conclusions

This chapter has presented an extremely brief outline of geography during its 'modern period', since the focus of the book is on the ensuing 'contemporary period'. Three paradigms have been identified, although deeper analysis may well indicate more coexisting during any one time (see Taylor, 1937). All three lasted into the contemporary period, although one, the regional paradigm, was dominant in the years before and just after the Second World War. Its main focus was on areal differentiation, on the varying character of the earth's surface (basically, the inhabited parts of the earth's surface), and its picture of that variation was built up out of parallel topical studies of different aspects of the physical and human patterns observed. By the 1950s, initially in America and then in Britain too, disillusionment with the empiricist philosophy of the regional paradigm was growing. Slowly the topical specialisms came to greater dominance and the regional synthesis was ignored: eventually, as outlined in the next chapter, a full revolution was launched with the presentation and ultimate wide acceptance of a new paradigm.

3

Growth of systematic studies and the adoption of 'scientific method'

Dating the origin of a change in the orientation of a discipline, or even a part of it, is difficult. Several pieces which contain the kernel of the new ideas can usually be found in its literature, but often these are derivative of the earlier teachings of others, whose views are never published. Further, it is possible for a change to emanate contemporaneously from several separate though usually not entirely independent nodes, as various iconoclasts introduce stimuli to change. An attempt to locate the first stirrings against the regional paradigm in human geography would be a futile exercise, therefore, and is not attempted here. Instead, the present chapter isolates what appear to have been the most important and influential statements published by geographers, and traces their impact on the geographical community of scholars.

As pointed out in Chapter 1, change within a discipline involves both dissatisfaction with the existing paradigm(s) and the preparation of an acceptable alternative. The existence of the former was spelled out by Freeman (1961), who noted that 'disapointment with the work of regional geographers has led many to wonder if the regional approach can ever be academically satisfying and to turn to specialization or some systematic branch of the subject' (p. 141). He suggested three reasons for such disappointment. The first was that so much regional classification was naive, particularly at the large scale, where generalizations, such as Herbertson's world climatic regions, were found on detailed investigation to contain too many discrepancies. The second, and perhaps most important to many people, was the 'weary succession' of physical and human activity 'facts' which characterized so much regional writing (though not all, as exemplified by James's, 1942, *Latin America*): 'The trouble has perhaps been that many regional geographers have tried to include too much' (p. 143). Thirdly, he claimed that the model of regional writing, derived from work on the French *pays*, suggested that the whole of the earth's surface could be divided into such clear regions, each with its own character: that this proved to be not so was reflected by many pedestrian studies of areas lacking such personality.

Whereas Freeman focused on the failings of regional geography as practised, a case made in the United States during the late 1940s and early 1950s was that the insistence of the primacy of regional geography was undermining the associated systematic studies. This was put forcefully by Ackerman (1945) in a paper reporting on his experience of working in the wartime intelligence services. He identified two major failings of professional geographers involved in such activities: their inability to handle foreign languages, and the weakness of their topical specialisms. Regarding the latter, he criticized much of the geographical work of the preceding quarter of a century as having been conducted by scholars who were 'more or less amateurs in the subject on which they published' (p. 124), so that when they were called upon to provide intelligence material for wartime interpretation what the geographers produced was extremely thin in its content. Regional geographers could provide no more than a superficial analysis, and the division of labour within the discipline whereby people specialized on different areas of the earth was both inefficient and ineffective. (Gould, 1979, p. 140, calls the geography of the fifty years prior to 1950 'bumbling amateurism and antiquarianism'.)

Ackerman suggested that rectification of this major deficiency in geographical work required much more research and training in the systematic specialisms: this would not be contrary to the philosophy of the subject which gave primacy to the regional synthesis, he claimed, since more detailed systematic studies would lead to greater depth in regional interpretations. There is little evidence that his paper had an immediate impact, however, and the publications of American geographers over the next few years, including the abstracts of the papers presented at the annual conferences of the Association of American Geographers, indicated no major shift in the orientation of academic work with the return to post-war 'normality': two of the few exceptions are the abstracts of papers presented by Garrison and McCarty at Cleveland in 1953, which were clearly based on a different methodology to that widely used (see below, p. 61). The systematic fields had undoubtedly been gaining in importance prior to Ackerman's statement, and continued to do so, as indicated by the extent of their treatment in the review volume edited by James and Jones (1954). But it wa not until the mid-1950s that this volumetric change in the substance of geographical research was matched by any widespread changes in its methodology and philosophy.

Schaefer's paper and the response

As it was in the United States that Hartshorne published his major statement of the regional paradigm, and as it was there, rather than in Britain, that philosophy and methodology were apparently debated most earnestly, it is perhaps not surprising that the revolution against the regional paradigm originated on that side of Atlantic. One of the first shots was a paper by Schaefer (1953) – which was published posthumously – that is often referred to by those who seek the origins of the 'quantitative and theoretical revolutions'. Schaefer was originally an economist: he joined the group of geographers teaching in the economics department at the University of Iowa after his escape from Nazi Germany (Bunge, 1979).

Schaefer pointed out that his paper was the first to challenge Hartshorne's presentation and interpretation of the works of Hettner and others, and it was published fourteen years after Hartshorne's monograph. His intent was to criticize the 'exceptionalist' claims made for regional geography, and to present the case for geography adopting the philosophy and methods of the positivist school of science. His first task, then, was to outline the nature of a science and to define the peculiar characteristics of geography as a social science. He argued that to claim that geography was the integrating science which put together the findings of the individual systematic sciences was arrogant, and that in any case its products were 'somewhat lacking in . . . startlingly new and deeper insights' (p. 227). A science is characterized by its explanations, and explanations require laws:

> To explain the phenomena one has described means always to recognize them as instances of laws (p. 227).

In geography, according to Schaefer, the major regularities which are described refer to spatial patterns:

> Hence geography has to be conceived as the science concerned with the formulation of the laws governing the spatial distribution of certain features on the surface of the earth (p. 227)

and it is these spatial arrangements of phenomena, and not the phenomena themselves, about which geographers should be seeking to make law-like statements. Geographical procedures would then not differ from those employed in the other sciences, both natural and soical: observation would lead to a hypothesis – about the interrelationship between two spatial patterns, for example – and this would be tested against large numbers of cases, to provide the material for a law if it were thereby verified.

The argument against this definition of geography as the science of spatial arrangements – the implicit argument in Hartshorne's work – Schaefer termed exceptionalist. It claims that geography does not share the methodology of other sciences because of the peculiar nature of its subject matter – the study of unique places, or regions. Using analogies from physics and economics, Schaefer argued that geography is not peculiar in its focus on unique phenomena; all sciences deal with unique events which can only be accounted for by an integration of laws from various systematic sciences, but this does not prevent – although it undoubtedly makes more difficult – the development of those laws.

> It is, therefore, absurd to maintain that the geographers are distinguished among the scientists through the integration of heterogeneous phenomena which they achieve. There is nothing extraordinary about geography in that respect (p. 231).

In the second part of his paper, Schaefer traced the exceptionalist view in geography back to an analogy drawn by Kant (1724–1804) between geography and history, an analogy repeated by Hettner and by Hartshorne (see above, p. 43). He quoted (p. 233) from Kant's *Physische Geographie* (Vol. I, p. 8) that 'Geography and history together fill up the entire area of our perception: geography that of space and history that of time'. But when Kant was working, Schaefer claims, history and geography were cosmologies, not sciences, and a cosmology is 'not rational science but at best thoughtful contemplation of the universe' (p. 332). Hettner, however, followed Kant's views and developed geography as a cosmology, arguing that both history and geography deal with the unique, and thus do not apply the methods of science. Schaefer argued that this is a false position, for in explaining what happened at a certain time period historians must integrate the laws of the social sciences. Time periods, like places, are undoubtedly unique assemblages of phenomena, but this does not preclude the use of laws in unravelling and explaining them. History and geography can both be sciences for

> What scientists do is . . . *They apply to each concrete situation jointly all the laws that involve the variables they have reason to believe are relevant* (p. 239).

Hartshorne argued that law-seeking is not a part of geography. According to Schaefer, however, Hartshorne disregarded one aspect of Hettner's writing which was nomothetic in its orientation, and in doing this he to some extent misled American geographers. (Muller-Wille, 1978, p. 55 claims that Hettner predated Christaller in the development of ideas regarding central place theory; Hartshorne presumably disregarded this too for he made no reference to the paper by Hettner cited by Muller-Wille.)

The final part of Schaefer's paper reviews some of the problems of applying his nomothetic (law-producing) philosophy to geography as a spatial, social science. He recognizes, for example, the problems of experimentation and of quantification, and suggests a methodology based on cartographic correlations. A major point concerns the difference between laws produced in geography and those from other, 'maturer' social sciences. The former are morphological; the latter are process: in order fully to comprehend the assemblages of the phenomena described in geographers' morphological laws, therefore, it is necessary to derive process laws from other social sciences, a procedure which requires team work (the last point was made also by Ackerman). Geography, then, according to Schaefer, is the source of the laws on location, which may be used to differentiate the regions of the earth's surface.

Hartshorne's response
Schaefer's paper did not produce much direct reaction, in print, despite later claims that it was a major stimulus to work in the genre which he proposed (Bunge, 1962). It did, however, draw considerable response from Hartshorne, in the form first of a letter to the editor of the *Annals* (Hartshorne, 1954b) and later three substantive pieces (Hartshorne, 1955, 1958, 1959): the last of the latter was another major book which, although probably not as

influential as the 1939 volume, showed the continued importance of Hartshorne to American geographers as an interpreter of their subject's methodology and philosophy.

The purpose of Hartshorne's first (1955) paper (which subsumed the earlier letter) was to indicate the many flaws which he identified in Schaefer's
scholarship (see also Gregory, 1978a, p. 31). He begins with a further discussion of the *mores* of methodological debate (Hartshorne, 1948): most of
the paper was organized to illustrate that Schaefer was limited in his references, drew unsupportable conclusions, and misrepresented the views of
others, so that 'In every paragraph, in nearly every sentence of this third
section, there is serious falsification, either by commission or by omission, of
the views of the writer discussed' (p. 236). (It should be noted that this statement refers to the third part of Schaefer's paper, which focused on the work
of Hettner.) In more general terms Hartshorne claimed that Schaefer's
paper 'ignores the normal standards of critical scholarship and in effect
offers nothing more than personal opinion, thinly disguised as literary and
historical analysis' (p. 244). Since Hartshorne himself (1959, p. 8) was a
strong believer that 'geography is what geographers have made it', to him all
methodological and philosophical statements should be based on a close and
careful analysis of the published works of others.

Although most of this paper is concerned to examine the nature of
Schaefer's 'evidence', in the final section Hartshorne turned to an examination of the anti-exceptionalist argument. He pointed out (p. 237) that in
coming to the conclusion that geography should take process laws from the
systematic sciences and use them to produce morphological laws, Schaefer
came very close in effect to preaching the sort of exceptionalist claim that he
sought to destroy. It could be argued, therefore, that Schaefer's critique 'is a
total fraud' (p. 237). Schaefer's position is summed up as 'geography must
be a science, science is the search for laws, and all phenomena of nature and
human life are subject to such laws and completely determinable by them'
(p. 242). Such scientific determinism is opposed to the summary of what
geographers do as set out in *The Nature of Geography*, which has in any case
been treated in a most cavalier way by Schaefer.

In his second paper, Hartshorne (1958) addressed Schaefer's claim that
Kant was the source of the exceptionalist view. Literary analysis suggests
that both Humbolt and Hettner reached the same view independently, being
unaware of Kant's views when they were writing. May (1970, p. 9) suggests
that both Hartshorne and Schaefer could have misunderstood Kant's conception of a science, however, and of the role of geography as a science,
although he confirms Hartshorne's dismissal of Schaefer's interpretation of
the source of Kant's ideas (see the later exchange between Hartshorne, 1972
and May, 1972).

The third and most substantial piece in Hartshorne's rebuttal of
Schaefer's argument was a monograph (Hartshore, 1959) entitled *Perspective on the Nature of Geography*, the production of which was stimulated

by Schaefer – and by requests from colleagues that he respond in detail to Schaefer's argument – but which was also used as a vehicle for a discussion of a wide range of other issues raised during the two decades since the publication of his original statement (Hartshorne, 1939). He organized the discussion in a framework of ten separate questions/topics: the aim was to provide a methodology by which geography could meet its need for 'new conceptual approaches and more effective ways of measuring the interrelationships of phenomena' (p. 9), which could only develop out of an understanding and acceptance of the subject's 'essential character'.

The first set of questions was concerned with the meaning of areal differentiation, with the definition of the earth's surface, with a discussion of the peculiar geographical interest in the integration of phenomena in 'the total reality [that] is there for study, and geography is the name of the section of empirical knowledge which has always been called upon to study that reality' (p. 33), and with the determination of what is significant for geographical study; it led Hartshorne to the definition that 'geography is that discipline that seeks *to describe and interpret the variable character from place to place of the earth as the world of man*' (p. 47). He considered that human and natural factors do not have to be identified separately – any prior insistence on this was a function of the arguments of environmental determinists – and that a division into human and physical geography is unfortunate, because it limits the range of possible integrations in the study of reality.

Turning to the study of temporal processes, Hartshorne argued that geographers need only study proximate genesis, since it is classification by form of appearance rather than by provenance which is important for the geographical investigation of areal differentiation: since most landforms are stable, or virtually so, from the point of view of man, for example, the study of their change is irrelevant to the aims of geography. According to this argument (see also p. 48), geomorphology, insofar as it is the study of landform genesis, is not part of geography; the study of landforms is. With regard to cultural features in the landscape, Hartshorne made an important distinction between expository description and explanatory description:

> geography is primarily concerned to describe . . . the variable character of areas as formed by existing features in interrelationships . . . explanatory description of features in the past must be kept subordinate to the primary purpose (p. 99).

Thus historical geography should be the expository description of the historical present 'but the purpose of such dips into the past is not to trace developments or seek origins but to facilitate comprehension of the present' (p. 106); studies of causal development and genesis are the prerogative of the systematic sciences.

In attempting an answer to the question 'Is geography divided between systematic and regional geography?' Hartshorne developed a position different from that which he took in *The Nature of Geography*. Thus in 1959 he accepted that studies of interrelationships could be arranged along a continuum, 'from those which analyse the most elementary complexes in areal variation over the world to those which analyse the most complex integrations in areal variation within small areas' (p. 121). The former are called topical studies and the latter regional studies, but whereas

> every truly geographical study involves the use of both the topical and the regional approach (p. 122)

there is no argument that one is superior over the other, as being that to which all geographers should aspire. In this presentation, therefore, Hartshorne somewhat downgraded the regional synthesis from its earlier centrality, in his view, in the geographical enterprise.

With regard to the important question raised by Schaefer's paper – 'Does geography seek to formulate scientific laws or to describe individual cases?' – Hartshorne argued for the latter, largely by pointing out the difficulties of establishing such laws through geographical investigations. Scientific laws must be based on large numbers of cases, but geographers study complex integrations in unique places; scientific laws can best be established in laboratory experiments which allow only a few independent variables to vary, but such work is impossible in geography; interpretation requires skills in the systematic sciences which are beyond the capability of geographers; scientific laws suggest some kind of determinism, but this is inappropriate to the human motivations which are in part the causes of landscape variations: for these reasons, the search for laws is irrelevant to geography. But laws are not the only means to the scientific end of comprehending reality in any case: instead

> Geography seeks (1) on the basis of empirical observation as independent as possible of the person of the observer, to describe phenomena with the maximum degree of accuracy and certainty; (2) on this basis, to classify the phenomena, as far as reality permits, in terms of generic concepts or universals; (3) through rational consideration of the facts thus secured and by logical processes of analysis and synthesis, including the construction and use wherever possible of general principles or laws of generic relationships, to attain the maximum comprehension of the scientific interrelationships of phenomena; and (4) to arrange these findings in orderly systems so that what is known leads directly to the margin of the unknown (pp. 169–70),

which, he says, is a perfectly respectable scientific goal. (It is very similar to the overall goal of positivist work – p. 71 – and is the reason why several commentators see very little difference in ends, if not means, between Hartshorne's work and that of the spatial scientists.)

Finally, in discussing geography's position within the classification of sciences, Hartshorne returned to the Hettnerian analogy of geography as a chorological science with history as a chronological science. This is valid, he argues, because it describes the way in which geographers have worked, on both topical and regional subjects, with reference to interrelationships and integrations within areas. (This view was revived by Harris, 1971.)

Reconciliations?

The last paragraph indicates the major basis of the methodological and philosophical difference between himself and Schaefer. Hartshorne's was a positive view of geography: geography is what geographers have made it. Schaefer's view, on the other hand, was a normative one, of what geography should be, irrespective of what it had been. Over the next decade after Hartshorne published his *Perspective* it was Schaefer's view which very largely prevailed, on both sides of the Atlantic, although the extent of Schaefer's personal influence based on his 1953 paper was probably very slight and the real iconoclasts of the 'revolution' were those discussed in the next section. (Indeed in Britain, although Hartshorne's two books were clearly widely read and referenced, Schaefer's paper was not. It receives no mention in Freeman's (1961; 1980a) books, none in Chorley and Haggett's (1965b) trail-blazing *Frontiers in Geographical Teaching*, and only one in their major (Chorley and Haggett, 1967) *Models in Geography* – in the chapter by Stoddart.) Thus it is not surprising that relatively little attention has been paid elsewhere in the geographical literature to the Schaefer/Hartshorne debate (Gregory, 1978a, p. 32).

An attempt at reconciling the views of the two protagonists, and to suggest that they were not so antagonistic in their views as they themselves suggested, has been provided by Guelke (1977a, 1978; see also Gregory, 1978a, p. 31 and Entrikin, 1981). He shows that in general terms Hartshorne was very much a supporter of the scientific method as defined by the positivists (see below, p. 71), but that he created his own problems regarding the application of this method in geography becasue of his view on uniqueness. Schaefer, on the other hand, accepted the full positivist position, and showed that uniqueness was a general problem of science, and not a peculiar characteristic of geography. Thus

> In extending the idea of uniqueness to everything, Schaefer effectively removed a major logical objection to the possibility of a law-seeking geography and demonstrated that Hartshorne's view of uniqueness as a special problem was untenable for anyone who accepted the scientific model of explanation (Guelke, 1977a, p. 380),

and Hartshorne's distinction between idiographic and nomothetic approaches was misleading. Both Hartshorne and Schaefer ignored the possibility of geographers being major 'law-consumers', however; to Hartshorne, the alternatives were either law-making or the description of unique places, whilst to Schaefer geographers had to develop morphological laws, and ignore the interest in process laws which characterizes the systematic sciences.

According to Guelke (1977a, p. 348), when Schaefer insisted on the need for geographers to develop laws 'he created a major crisis within the discipline'. Whether Schaefer himself was responsible for the crisis is doubtful, as the next section suggests. There is no doubt, however, that within about a decade of Schaefer's paper being published, most human geographers, and certainly those of the youngest generation currently within the profession, had adopted at least part of his manifesto, with the growing concern for quantification and law-making. They were presented with a choice between such activity and the sort of contemplation of the unique advocated by Hartshorne. As Guelke (1977a, p. 385) points out, 'Not surprisingly, most geographers opted for geography as a law-seeking science' because (Guelke, 1978, p. 45) by then:

> Universities were expected to produce problem-solvers or social-technologists to run increasingly complex economies, and geographers were not slow in adopting new positions appropriate to the new conditions. Statistics and models were ideal tools for monitoring and planning in complex industrial societies. The work of the new geographers, however, often lacked a truly intellectual dimension. Many geographers were asking: 'Are our methods rigorous?', 'What are the planning implications of this model?', and not 'How much insight does this study give us?', 'Is my understanding of this phenomenon enhanced?', 'Does this study contribute to geography?'. The last-mentioned question was considered of little consequence. Yet it should have been asked, because one of the weaknesses of the new geography was a lack of coherence.

Developments in systematic geography in the United States

Whether because of, or independently of, the statements of Ackerman (1945), Schaefer (1953), and Ullman (1953), it is clear that during the 1950s systematic studies became much more important in the research and teaching of American geographers. This did not mean a departure from Hartshorne's views, since by 1959 he no longer gave primacy to regional studies, but the trend towards the scientific method proposed by Schaefer did mark a break with the Hartshornian tradition.

The growing popularity of topical specialisms is shown by the review chapters in the collection edited by James and Jones (1954) and by the journal literature of the 1950s. Very few of the investigations reported were aimed at the generation of laws in any sense, however: indeed some could almost be categorized under the exploration paradigm, in that their major purpose seemed to be the provision of new factual material about, for example, the changing distribution of Soviet pulp and paper industries (Rodgers, 1955): such work is best described an empiricist – it lets 'the facts' speak for themselves – and is positivist in its use, eventually, in the search for generalizations, perhaps by other workers.

Fundamental to the progress of science in the positivist mould espoused by Schaefer is the development of theory. Several of the reviews in the James and Jones (1954a) volume refer to what is in one place termed location theory (Harris, 1954, p. 299), but very few examples are cited of empirical investigations related to that body of theory. The chapter on urban geography, for example, cites all of the seminal pieces on central place theory, such as Ullman's (1941) original paper, and devotes two pages (Mayer, 1954, pp. 152–63) to the three 'models' of intra-urban spatial patterns which had been reviewed a decade previously by Harris and Ullman (1945), but there is not a single reference to any work done by geographers in the context of those models. Thus although there were some precedents in the literature, in general very little work had been done by goegraphers, prior to the mid-1950s, which followed the dictates of the 'scientific approach'.

Once a new idea gains circulation through the professional journals it is available to be taken up by all. Nevertheless, development of the idea is usually concentrated in a few places only, where the pioneer teachers encourage their students – in the American case, particularly their graduate students – to conduct research within the new framework. Thus most of the methodological changes in systematic studies in geography during the 1950s can be traced back to a few centres in the United States, and it is the work conducted at those centres which is discussed here. It should be stressed at the outset that the changes to be outlined were largely concerned with method, and that their scientific underpinning was stressed very little, although law-seeking was the clear goal. Certainly it was methods which dominated the literature, of both the pros and cons; many of the early contributions of the former groups were in relatively fugitive, departmental publications, presumably because of difficulties in getting such 'new' material accepted by the journals.

The Iowa school

Although Schaefer was at Iowa until his death in 1953, he was not the major influence on the developments which occurred there among the geographers who were, for a number of years, part of the economics department and thus open to the views and approaches of their peers in that more 'mature' social science. The leader of this group of geographers was Harold McCarty, author of a major text on American economic geography (McCarty, 1940). Associated with him were included J. C. Hook, D. S. Knos, H. A. Stafford, and, later, J. B. Lindberg, E. N. Thomas and L. J. King (McCarty, 1979; King, 1979a).

The intent of McCarty and his co-workers was to establish the degree of correspondence between two or more geographical patterns, akin to the morphological laws of accordance discussed by Schaefer. (Interestingly, none of their publications refers to Schaefer's paper, although they do refer to the works of, and assistance given by, Gustav Bergmann, a positivist of the Vienna School who also strongly influenced Schaefer and read the proofs of his 1953 paper: Davies, 1972, p. 134; King, 1979a.) These laws were to be embedded in a theory; thus (McCarty, 1954, p. 96):

If we are to accept the idea that economic geography is becoming the branch of human knowledge whose function is to account for the location of economic activities on the various portions of the earth's surface, it seems reasonable to expect the discipline to develop a body of theory to facilitate the performance of this task.

Such a theory could be either topically or areally focused, and in its early stages of development would probably be restricted both in its areal coverage and in the topics whose spatial interrelationships it considered.

The purpose of theory is to provide explanations, and McCarty recognized two sorts of explanation. The first is based on a search for the cause of observed locational patterns, but:

the search for causes can never produce an adequate body of theory for use in economic geography. . . . Variables became so numerous that they were not manageable, and, in consequence, solutions to locational problems were not obtainable (p. 96).

The second, and preferred, type focuses on associations:

Its proponents take the pragmatic view that if one knew that two phenomena always appear together in space and never appear independently, the needs of geographic science would be satisfied, and there would be scant additional virtue in knowing that the location of one phenomenon caused the location of another (p. 97).

Such laws of association are built up in a series of stages, which begins with a statement of the problem and of the necessary operational definitions, and proceeds through the measurement of the phenomena (with attendant problems of sampling in time and space) to a statement of the findings, in tabular or graphical form. These three descriptive stages precede analysis which seeks out correlations between the distributions of phenomena:

the nub of problem of research procedure seems to lie in finding the best techniques for discovering a, b and c in the 'where a, b, c, there x' hypothesis in order to give direction to the analysis. But where shall we search for its components? . . . One source of . . . clues lies in the findings of the systematic sciences. The other source lies in the observations of trained workers in the field or in the library (p. 100).

Thus geography, in its search for morphological laws, is to a considerable extent a consumer of the laws of other disciplines. These laws may be theoretically rather than empirically derived. According to the causal or process approach to explanation:

Models may be created showing optimal locations for any type of economic activity for which adequate cost data may be obtained. These models may then be used (as hypotheses) for the comparison of hypothetical locations with actual locations. Divergences of pattern may then be noted and the hypothesis altered to allow for them (often by inclusion of factors not ordinarily associated with monetary costs). Ultimately the hypothesis becomes generally applicable and thus takes on the status of a principle (McCarty, 1953, p. 184).

This statement, although not referenced as such, very faithfully reflects the views of positivists, and also of Popper and Lakatos, as to how science progresses by the continual modification of its hypotheses, so as better to represent reality.

In their major demonstration of this procedure in operation, McCarty *et al.* (1956) discussed several statistical procedures for measuring spatial association and adopted the now well-known technique of multiple regression and correlation which had been used previously, among geographers, by Rose (1936) and Weaver (1943) – both apparently as a product of contacts with agricultural economists. Their empirical context was similarity in the location patterns of certain manufacturing industries in the United States and in Japan: other studies by the group included Hook's (1955) on rural population densities, Knos's (1968) on intra-urban land-value patterns, and King's (1961) on the spacing of urban settlements. Thomas (1960) had used similar procedures in his study of population growth in suburban Chicago, presented as a PhD thesis to Northwestern University, and he extended the methodology with a paper on the use of residuals from regression for both identifying where the laws of association do not apply fully and suggesting further hypotheses for areal associations. (This last paper developed on an earlier one by McCarty (1952) which was not widely circulated.) Later, McCarty (1958) expressed some doubts about the statistical validity of the procedure, but the method that he and his associates pioneered, with its focus on the testing of simple hypotheses derived either from observation or from theoretical deductions, became the model for much research in the ensuing decades.

Wisconsin

The Department of Geography at the University of Wisconsin, Madison, has a long tradition of research with a quantitative bent. Notable among its early products was the PhD thesis of John Weaver on the geography of American barley production, which included a major section (published in his 1943 paper with no supporting methodological argument) using multiple correlation and regression to identify the influences of climatic variables on barley yields. (Weaver later taught at the University of Minnesota, where he developed a widely adopted statistical procedure (Weaver, 1954) for the definition of agricultural regions.) Other work at Madison focused on the measurement of population patterns (e.g. Alexander and Zahorchak, 1943). A combination of these two interests was furthered by a group led by A. H. Robinson, whose main interests have been in cartography, and cartographic correlations were apparently introduced to him by his research supervisor at Ohio State University, Guy Harold Smith (Brown, 1978); Robinson was assisted by R. A. Bryson, of the Department of Meteorology, who was a source of statistical ideas and expertise.

(Cartographic work was for a long time called 'mathematical geography' by some. It also developed statistical interests: see Blumenstock, 1953.)

Robinson's concern was to develop statistical methods of map comparison, as indicated by the title of his early paper – 'A method for describing quantitatively the correspondence of geographical distributions' (Robinson and Bryson, 1957). As with the work done at Iowa, the lead of Rose and of Weaver was followed with the adoption of correlation and regression procedures. Particular attention was paid to the problems of representing areal data by points (Robinson *et al.*, 1961) and of using correlation methods in the comparison of isarithmic maps (Robinson, 1962). Like McCarty (1958), Robinson was aware of difficulties in applying classical statistical procedures to areal data, and he proposed a procedure to circumvent one of these (Robinson, 1956): Thomas and Anderson (1965) later found this proposal wanting, as it dealt with a special case only and not with the more general problems. Interestingly, though, the main early work on this topic was published by a group of sociologists, under the title of *Statistical Geography* (Duncan, Cuzzort and Duncan, 1961: perhaps even more interestingly, this work was virtually ignored by geographers).

W. L. Garrison and the Washington school

By far the largest volume of work in the spirit of Schaefer's and McCarty's proposals that was published during the 1950s came from the University of Washington, Seattle. The leader of the group of workers there was W. L. Garrison, whose PhD was from Northwestern University, where he was associated with Thomas (the two returned to Northwestern in the early 1960s; Taaffe, 1979), and who, according to Bunge (1966, p. ix), was influenced by Schaefer's paper, although the dates of his earliest publications indicate that he was involved in applying the positivist method to systematic studies in human geography before 1953. Also involved was E. L. Ullman, who moved to Seattle in 1951 (Harris, 1977), and who had already done pioneering research into urban location patterns and transport geography, and general support to Garrison's efforts was given by the departmental chairman, G. Donald Hudson. A large group of graduate students worked with Garrison and several became leaders in the new methodology during the subsequent decade, including B. J. L. Berry, W. Bunge, M. F. Dacey, A. Getis, D. F. Marble, R. L. Morrill, J. D. Nystuen, and W. R. Tobler (Garrison, 1979). The group also benefited from a visit to Seattle by the Swedish geographer, Torsten Hägerstrand, who was developing methods of generalizing spatial patterns and processes (see below, p. 96), and from Garrison's contacts with the Business School and the Engineering department at Seattle (Halvorson and Stave, 1978).

Garrison and his co-workers had catholic interests in urban and economic geography. Much of their work was grounded in theory which they gleaned from other disciplines – notably economics – and they directed their efforts towards both testing those theories and applying them to problems of planning. (Another staff member at Seattle, M. E. Marts, had a major interest in regional planning.) In developing testable theoretical statements they displayed a much stronger mathematical base than was the case at Iowa and Madison. They also searched widely for statistical tests which were relevant to their investigation of point and line patterns – the biological sciences provided several of those which they adopted, such as nearest-neighbour analysis of point patterns (Dacey, 1962), and others, used for grouping and classifying, were derived from psychology (Berry, 1968). Garrison's (1956a, p. 429) view was that 'there is ample evidence that present tools are adequate to our present state of development. No type of problem has been proposed that could not be treated with available tools,' which was in contradiction to an earlier claim by Reynolds (1956), although he was critical of how some standard techniques had been used (Garrison, 1956b). The dominant thrust of the group's work, therefore, involved the derivation from other systematic sciences of relevant normative theories, mathematical methods and statistical procedures with which to develop morphological laws.

The wealth of the work done at Seattle is best illustrated by a few of their major publications. Garrison contributed an important three-part review article, for example (Garrison, 1959a, 1959b, 1960a), on the state of location theory. The first part comprised a review of six recent books – none of them by geographers – devoted to the question 'What determines the spatial arrangement (structure, pattern, or location) of economic activity?' (Garrison, 1959a, p. 232). Each of the books incorporated locational considerations into traditional economic analysis, and Garrison concluded that this offered valuable economic insights to traditional geographical problems; for geographers to advance their locational analyses and theories they would have to pay much more attention to the work of economists.

Among the Washington group, central place theory was the dominant location theory on which they worked. This had several independent origins, as Ullman (1941) had indicated (see also Harris, 1977, and Freeman, 1961, p. 201, who notes that findings akin to those in central place theory were reported by the 1851 Census Commissioners of Great Britain). It was Christaller's (1966) thesis which attracted most attention, however (see Muller-Wille, 1978). Working in Germany in the 1930s, Christaller developed ideas regarding the ideal distribution of settlements of different sizes acting as the marketing centres of functional regions, within the constraints of assumptions relating to the physical environment and the goals of both entrepreneurs and customers: a translation became available in the late 1950s, and was published in 1966. Studies of functional regions were not novel, of course (see above, p. 41), and geographers had already attempted to test Christaller's notions regarding a hierarchical organization of settlements distributed on a hexagonal lattice (e.g. Brush, 1953); related work, on a more inductive base, on 'principles of areal organization', was reported by Philbrick (1957). Dacey was concerned to make the analysis of these spatial hierarchies more rigorous (Dacey, 1962) and Berry's research focused on the central place aspects of the settlement pattern north of Seattle and on the retail centres in the city of Spokane (Berry and Garrison, 1958a, 1958b; Berry, 1959a).

The second part of Garrison's (1959b) review article dealt with possible geographical applications of the mathematical procedures of linear programming, which produce the optimal solutions to problems of resource allocation in constrained situations. In this, he illustrated how the procedures of neoclassical economic analysis could be adapted in order to investigate ideal solutions to the problems of where to locate economic activities and how to organize flows of goods. Six such problems which could be treated by linear programming were identified:

1   the transportation problem, which takes a set of points, some with a given supply of a good and some with a given demand for it, plus costs of movement, and determines the most efficient flow pattern of the good from supply to demand points which minimizes the expenditure on transport;
2   the spatial price-equilibrium problem, which takes the same information as the transportation problem, but is used to determine prices as well as flows;
3   the warehouse-location problem, which determines the best location for a set of supply points, given a geography of demand;
4   the industrial-location problem, which determines the optimum location for factories from knowledge of the sources of their raw materials and the destinations for their products;
5   the interdependencies problem, which locates linked plants so as to maximize their joint profits; and
6   the boundary-drawing problem, which determines the most efficient set of boundaries (i.e. that which minimizes total expenditure on transport) for, for example, school catchment areas.

The purpose of these analyses, if they are being used to investigate actual patterns and not as the bases for future plans, is, as Lösch (1954) put it, to see whether reality is rational, whether decision-makers have acted in ways that would produce the most efficient solutions, where efficiency is defined as cost-minimization, particularly transport-cost minimization. Investigations by the Washington group in this context included studies of interregional trade (Morrill and Garrison, 1960) and the optimal location, by regions, of agricultural activities in the United States (Garrison and Marble, 1957).

In the final part of his review article, Garrison (1960a) dealt with four further books on locational analysis, which were empirical in orientation and shared a common interest in the agglomeration economies reaped by industrial clusters. Several topics and techniques were discussed, such as the use of input-output matrices to represent industrial systems, and Garrison concluded by stressing the need for geographers to investigate location patterns as systems of interrelated activities, a topic which is taken up again in the next chapter.

The empirical work with a planning orientation undertaken by the Seattle group is illustrated by one of the outputs of Garrison's large study of the impact of highway developments on land use and other spatial patterns (Garrison *et al.*, 1959). This book includes four studies: Berry's on the spatial pattern of central places within urban areas; Marble's on the residential pattern of the city (as indexed by property values) and relationships between household characteristics – including location – and their movement patterns; Nystuen's on movements by customers to central places; and Morrill's on the locations of physicians' offices, both actual and the most efficient. In addition, Garrison himself worked on the accessibility impacts of highway improvements, and devised indices of accessibility based on graph theory (Garrison, 1960b: the work was continued by Kansky, 1963, after Garrison and Berry moved to work in the Chicago area). He also used the simulation procedures that had been developed by Hägerstrand (1968) to investigate urban growth processes (Garrison, 1962), a topic taken much further by Morrill (1965). Berry (1958) extended his work on central places with a piece on planned shopping centres.

Somewhat separate from the work of the others in the group, although aligned with their general purpose, was Bunge's thesis, *Theoretical Geography* (1962, reprinted in enlarged form, 1966). This too displays a catholic view of geography, together with an acknowledged debt to Schaefer (Bunge worked at Iowa for a short period, and also at Madison). It is an extremely difficult book to summarize, but the basic theme is very clear: geography is the science of spatial relations and interrelations; geometry is the mathematics of space; hence geometry is the language of geography. Thus the early chapters are concerned with establishing geography's scientific credentials, in a debate with Hartshorne's published statements, especially those concerning uniqueness and predictability. As Lewis (1965) and other have also argued, Bunge claimed that Hartshorne confused uniqueness and singularity: he opposed Hartshorne's claim that geography cannot formulate laws because of its paucity of cases by arguing for even more general laws, and countered the argument that geographical phenomena are not predictable with the claim that science 'does not strive for complete accuracy but compromises its accuracy for generality' (p. 12).

Having established geography's scientific credibility to his satisfaction, Bunge then investigated its language. An intriguing discussion of cartography led him to conclude that descriptive mathematics is preferable to cartography as a more precise language. The remainder of the book looked at aspects of the substantive content of the science of geography, beginning with 'a general theory of movement' and then a chapter on central place theory:

> If it were not for the existence of central place theory, it would not be possible to be so emphatic about the existence of a theoretical geography . . . central place theory is geography's finest intellectual product (p. 133).

Problems of testing the theory were then discussed, showing, as did Getis (1963), the need for map transformations, and in the final chapter of the first edition the links between geography and geometry were made clear:

> Now that the science of space is maturing so rapidly, the mathematics of space – geometry – should be utilized with an efficiency never achieved by other sciences (p. 201).

The richness of the work done by this group during the mid and late 1950s continued in various locations after it broke up (only Ullman remained, although Morrill later joined the staff at Seattle). Most prolific and seminal has been Berry, not only in his original field of central place theory (Berry, 1967) but also over a very wide range of other topics in economic and social geography. Berry's work has always had a very strong empirical and utilitarian base, whereas Dacey has continued to work on the mathematical representation of spatial, especially point, patterns (e.g. Dacey, 1973). In total, the work of this group of scholars influenced the research and teaching of a whole generation of human geographers, throughout the world, and has undoubtedly been the most significant in the development of the first post-regional paradigm.

The social physics school
The work of this group was initiated and developed independently from that of any of the other three – its early publications preceded Schaefer's paper by more than a decade. The leader was J. Q. Stewart, an astronomer at Princeton University, who traced the origins of social physics in the work of a number of natural scientists who applied their methods to social data (Stewart, 1950). His own work apparently began when he noted certain regularities in various aspects of population distributions, regularities which were akin to the laws of physics, such as a tendency for the number of students attending a particular university to decline with increasing distance of their home addresses from its campus. From these observations he developed his ideas on social physics, which he defined (Stewart, 1956, p. 245) as:

> . . . that the dimensions of society are analogous to the physical dimensions and include numbers of people, distance, and time. Social physics deals with observations, processes and relations in these terms. The distinction between it and mathematical statistics is no more difficult to draw than for certain other phases of physics. The distinction between social physics and sociology is the avoidance of subjective descriptions in the former.

and he established a laboratory at Princeton to investigate the wide range of regularities which could be analysed in this context.

Stewart's ideas were introduced to geographers by a paper in the *Geographical Review* (Stewart, 1947). Four empirical rules were adduced there: the first, the rank-size rule for cities, showed that in the United States the population of a city multiplied by its rank (from 1 for the largest to n for the smallest), and standardized by a constant, equalled the population of the largest city, New York (Carroll, 1982); the second indicated that at various dates the number of cities in the country with populations exceeding 2500 was very closely related to the proportion of the population living in such places; the third showed that the distribution of a population could be described by the population potential at a series of points, in the same manner as the potential in a magnetic field is described in Newtonian physics; and the fourth illustrated a close relationship between this population potential and the density of rural population in the United States. From these regularities, Stewart claimed that

> There is no longer any excuse for anyone to ignore the fact that human beings, on the average and at least in certain circumstances, obey mathematical rules resembling in a general way some of the primitive 'laws' of physics (p. 485).

No reasons were given why this should be so (Curry, 1967b, p. 285, called this a 'deliberate shunning of plausible argument'): the rules were presented as empirical regularities which had some similarity to the basic laws of physics. Causal hypotheses were not even postulated, let alone tested.

One of Stewart's collaborators was William Warntz, a graduate of the University of Pennsylvania who was later employed by the American Geographical Society as a research associate, working on what he termed 'the investigation of distance as one of the basic dimensions of society' (Warntz, 1959b, p. 449). The wide range of empirical regularities which they observed (see Stewart and Warntz, 1958, 1959) was used to develop their concept of macrogeography (Warntz, 1959b, 1959c). Geographical work, Warntz claimed, was dominated by micro-studies:

> The tendency of American geographers to be preoccupied with the unique, the exceptional, the immediate, the microscopic, the demonstrably utilitarian, and often the obvious is at once a strength and a weakness (Warntz, 1959b, p. 447)

but

> the assembly of more and more area studies involving an increase in the quantity of detail does not mean *per se* a shift from the microscopic to the macroscopic (p. 449).

Thus geographers were in danger of being unable to perceive general patterns within their welter of local detail. To counter this, Stewart and Warntz suggested the search for 'regularities in the aggregate'. Stewart's concept of population potential was used to describe general distributions, and was shown to be related to a large number of other patterns in the economic and social geography of the United States. These findings, it was realized, were only empirical regularities, but they could be used as the basis for theory development (Stewart and Warntz, 1958, p. 172), for geography's need was theory which 'has as its aim the establishment and coordination of areal relations among observed phenomena. General laws are sought that will serve to unify the individual, apparently unique, isolated facts so laboriously collected' (Warntz, 1959b, p. 58). The approach to theory, then, was inductive (see Figure 3.1) rather than deductive, although there was a clear underlying belief in the importance of distance and accessibility as influences on individual behaviour. To Warntz, as to Garrison, Ullman, and others, the ideas of Christaller were seminal (see Bunge, 1968).

These macroscopic measures, particularly that of population potential, were used in a variety of contexts, as in Warntz's (1959a) *Towards a Geography of Price*, which established strong relationships between the prices of agricultural commodities in the USA and measures of supply and demand potential; Harris (1954b) and Pred (1965a) also used the potential measure in their studies of industrial-location patterns. The 'macrogeographers' also did much work on various distance-decay functions (see Chapter 4) and developed the early work of a Russian group on centrographic measures (Sviatlovsky and Eels, 1937; Neft, 1966).

These lines of work contrasted markedly with that of the other three groups reviewed here, in a variety of ways. First was the topic of scale; Stewart and Warntz perhaps conformed more than any others to Bunge's call for a scientific approach which aimed at a high level of generality. Second there was the nature of the approach to theory, for macrogeography was inductive in its search for regularity rather than testing deductive hypotheses. Finally, the analogies sought for human geography were in a natural science – physics – and not in the other social sciences.

Summary

The developments outlined in this section marked the beginning of major changes in the field of human geography, changes which were rapidly taken up by others, within and beyond the United Sates. Although the focus was on theory and measurement, and the development of 'geographical laws', in line with the general ethos of academia in the immediate post-war decades, to some extent the work did not deviate too far from Hartshorne's expanded definition of the nature of geography (see above, p. 55). The main difference between the new work, with its focus in the systematic studies, and the regional paradigm was the greater faith of geographers in their ability to produce laws, to work within the canons of accepted scientific method, and to move out of their self-imposed academic isolation (Ackerman, 1945, 1963).

The scientific method in geography

Whether or not the changes just described initiated a 'Kuhnian revolution' in human geography is a topic left to the last chapter of this book. It is clear, however, that they involved at least a major reorientation in the nature of geographical research. Perhaps surprisingly, this reorientation was focused on no programmatic statement. There was no published paper or book which provided a detailed outline of how research should be conducted in this new framework: Schaefer's paper said nothing about how geographical laws were to be stated and derived, and although McCarty and his co-workers discussed methods in both general (McCarty, 1954) and specific (McCarty *et al.*, 1956) contexts, they did not provide an overall programme. As Gregory (1978a, p. 47) puts it, 'geography has (with some notable exceptions) paid scant attention to its epistemological foundations'.

One piece was not central to the initial efforts, because it post-dated many of them, but it was widely quoted in the 1960s, as the new ideas spread. Ackerman's (1958) essay was entitled *Geography as a Fundamental Research Discipline* and was an analysis of research organization. He indicated that the ultimate goal was integration to provide a full comprehension of reality, and that with regard to current developments, 'If any one theme may be used to characterize this period, that theme would be one of illuminating covariant relations among earth features' (p. 7). As a science, even one which is eventually an idiographic science since it deals with unique places, geography, according to Ackerman, needed to strive for 'an increasingly nomothetic component'. Its fundamental research

> need not necessarily be law-giving. . . . Much fundamental research in geography has not been law-giving in the strict sense but it has been concerned with a high level of generalization, and it has given meaning to other research efforts which succeeded it. In this sense it has a block-building characteristic (p. 17).

Such fundamental research 'is likely to rest on quantification . . . accurate study depends on quantification' (p. 30) and should 'furnish a theoretical framework with capacity to illuminate actually observed distributional patterns and space relations' (p. 28).

Ackerman's essay was a clarion call for the development of theory, for the application of quantitative methods, and for a focus on laws and generalizations to form the building-blocks for further nomothetic research. But there was no detailed discussion of how such research should be undertaken: it was a clarion call without reference to detail. Seven years later, the report of a National Academy of Sciences-National Research Council (1965) committee on *The Science of Geography* discussed 'Geography's problem and method' with the statement that

> Geographers believe that correlations of spatial distributions, considered both statistically and dynamically, may be the most ready keys to understanding existing or developing life systems, social systems, or environmental changes. In the past . . . progress was gradual, however, because geographers were few, rigorous methods for analysing multivariate problems and systems concepts were developed only recently (p. 9).

Again, a general statement on orientation of research, but no detail on how research was to be conducted. And yet, in a paper published in 1963, Burton claimed that an intellectual revolution – the quantitative and theoretical revolution – had occurred in geography: 'The revolution is over, in that once-revolutionary ideas are now conventional' (p. 156). Something had

Figure 3.1   Two routes to scientific explanation. Source: Harvey (1969a, p. 34).

become conventional, but nobody had written a full formulation for the discipline of what that something was!

The scientific method

There is no suggestion here that the groups of researchers who proposed changes in the nature of human geography had no clear rationale for their work; indeed, those involved were undoubtedly very clear as to both means and ends (though this may not be the case with some of their disciples), but they did not discuss these in detail in print. Nor, if their citations in the published works are any lead, did they research deeply into the philosophy that they were adopting: the exceptions to this are the references to Bergmann in the Iowa group's papers and in Bunge's thesis. Texts on statistical and mathematical procedures were widely quoted, but the first major work on the philosophy of the 'new geography' was not published until 1969, in a book which received wide acclaim (Harvey, 1969a: a parallel, but briefer statement is Moss, 1970: note that Harvey was trained in Britain and his book was published there).

There are two routes to explanation, according to Harvey (Figure 3.1). The first, occasionally known as the 'Baconian' or inductive route, derives its generalizations out of observations: a pattern is observed and an explanation developed from and for it. This involves a dangerous form of generalizing from the particular case, however, because, as Moss (1970) argues, acceptance of the interpretations depends too much on the charisma of the scholar involved. So the preferred method is the second route in Figure 3.1. This also begins with an observer perceiving patterns in the world; he then formulates experiments, or some other kind of test, to prove the veracity of the explanations which he produced for those patterns. Only when his ideas have been tested successfully against data other than those from which they were derived can a generalization be produced.

Scientific knowledge, obtained via the second route, is 'a kind of controlled speculation' (Harvey, 1969a, p. 35), and it is such a procedure that an increasing number of human geographers sought to apply during the 1950s. The method, known as positivism, was developed by a group of philosophers working in Vienna during the 1920s and 1930s (Guelke, 1978). It is based on a conception of an objective world in which there is order waiting to be discovered. Because that order – the spatial patterns of variation and covariation in the case of geography – exists, it cannot be contaminated by the observer. A neutral observer, on the basis of either his observations or his reading of the research of others, will derive a hypothesis (a speculative law) about some aspect of reality and then test that hypothesis: verification of his hypothesis translates the speculative law into an accepted one.

A key concept in this philosophy is that laws must be proven through objective procedures, and not accepted simply because they seem plausible: as Bunge (1962, p. 3) puts it, 'the plausibility or intuitive reality of a theory is *not* a valid basis for judging a theory'. A valid law must predict certain patterns in the world, so that having developed an idea about those patterns, the researcher must formulate them into a testable *hypothesis* – 'a proposition whose truth or falsity is capable of being asserted' (Harvey, 1969a, p. 100). An experiment is then designed to test the hypothesis, data are collected, and the validity of the predictions evaluated.

If the results of the test do not match the predictions, then either the observations on which the hypothesis was based or the deductions from the works of others are thrown into doubt. There is thus negative feedback (Figure 3.1) and the image of the world has to be revised, creating a new hypothesis. (This in fact is the Popperian view that any hypothesis is found wanting by a single falsification. Harvey (1969a, p. 39) gives only eight lines to this view, however, preferring the more general one that only 'severe failure' – which he does not define – discredits an hypothesis totally: see, however, Moss, 1977; Bird, 1975; Petch and Haines-Young, 1979.) Clearly, if the researcher is a good observer and a logical thinker, such falsification of his hypotheses should be rare. If the test is successful, on the other hand, then the speculation of the hypothesis becomes an acceptable generalization. One successful test will not turn it into a *law*: replication on other data sets will be needed, since a law is supposed to be a universal.

According to Harvey (1969a, p. 105):

> A scientific law may be interpreted as a generalization which is empirically universally true, and one which is also an integral part of a theoretical system in which we have supreme confidence. Such a rigid interpretation would probably mean that scientific laws would be non-existent in all of the sciences. Scientists therefore relax their criteria to some degree in their practical application of the term.

After sufficient (undefined) successful tests, therefore, a hypothesis may be accorded law-like status, and is fed into a body of *theory*, which comprises a series of related laws. There are two types of statement within a full theory: the *axioms*, or givens, which are statements taken to be true, such as laws; and the deductions, or *theorems*, from those initial conditions, which are derived consequences from agreed facts – the next round of hypotheses. There is a positive feedback from the theory stage to the world view, therefore (Figure 3.1), so that the whole scientific enterprise, aimed at total explanation as Ackerman argued, is a circular procedure whereby the successes of one set of experiments become the building blocks for thinking about the next.

One stage in Figure 3.1 so far ignored is the *model*, a widely used term which has been given a variety of meanings (Chorley, 1964). There are two basic functions of a model: as a *representation of the real world*, such as a scale model, a map, a series of equations, and some other analogue (Morgan, 1967); and an *ideal type*, a representation of the world under certain constrained conditions. Both are used in the positivist method to operationalize a theory, as a guide to the derivation of testable hypotheses.

Quantification has a central function in this scientific method. Mathematics are particularly useful in the writing of models, as in the linear programming procedures adopted by Garrison. Relatively few geographers have strong backgrounds in mathematics, however (this was especially true in the 1950s), and so little work was done which involved representation of the real world as sets of equations. Instead the central role was given to statistics, used in hypothesis-testing. Two types of statistics are available: *descriptive statistics* can be used to represent a pattern or relationship; *inductive statistics* are used to make generalizations about a carefully defined population from a properly selected random sample of that population. Many geographical researchers confused the two. Inductive statistics use significance tests to show whether what has been observed in the sample probably also occurs in the parent population, so that if the data analysed are not a sample, such tests are irrelevant (some disagreement has been expressed over 'what is a sample?': see Meyer, 1972 and Court, 1972). Many geographers have used inductive methods in a descriptive manner, however, using the significance tests as measures of the validity of their findings.

The main attraction of statistics to many of the early adherents of the 'new geography' was their precision and lack of ambiguity – compared to the English language – in description. This was expressed by Cole (1969) who annotated a quotation from a well-known text (Stamp and Beaver, 1947, pp. 164–5): in this the text is Stamp and Beaver's and the annotations in parentheses, to shown the ambiguities, are Cole's:

> The present distribution of wheat cultivation in the British Isles (space) raises the conception of two different types of limit. Broadly speaking (vague), it may be said that the possible (vague) limits (limit) of cultivation of any crop are determined by geographical (vague), primarily by climatic, conditions. The limits so determined (how?) may be described (definition) as the ultimate (vague) or the geographical (vague) limits . . . (Cole, 1969, p. 160).

The full quotation, according to Cole (only part is reproduced here), is so full of ambiguities that it could refer to about one million million possible combinations of some forty counties and it is impossible to reconstruct a map from that description:

> the correlations suggested are so tentative and imprecise that they leave the reader still wondering why wheat is grown where it is. The application of a standard correlation procedure . . . in itself would give a more precise appreciation of the relationship (p. 162).

Similar views became widely held during the 1950s and 1960s, and quantification became the *sine qua non* of training in the new methods (LaValle, McConnell, and Brown, 1967).

The scientific method increasingly adopted by geographers is a procedure for testing ideas, therefore, but a highly formalized one, about which there has been a great deal of debate (Harvey, 1969a). Although many aspects of the method were used by geographers, their citations indicate relatively little research in depth into the full philosophy of positivism. (Positivism, as used here, refers to what in often known as 'scientific method'. It is embraced by the philosophy of logical positivism, which claims that only scientifically obtained knowledge is valid knowledge: Johnston, 1983b.)

Reactions to the scientific method

Despite (or perhaps because of) the lack of a clear programmatic statment of the 'new Theology' (Stamp, 1966, p. 18) until Harvey's (1969a) book, reactions to the developments were many and varied. (James, 1965, p. 35, called the debate 'continued, bitter and uncompromising warfare'.) Two related issues were the main foci of contention: whether quantification was sensible in geographical research, and whether law-making was possible. As Taylor (1976) points out, to some extent the debate was inter-generational, of the type discussed in Chapter 1 (p. 27): to some of the 'old guard' what was being proposed was just not geography and should be banished to some other corner of academia.

The quantification issue was the less important, and few spoke out against it in its entirety, although its extent was criticized. Thus Spate (1960a, p. 387) recognized that quantification is 'an essential element' and

> This is, like it or not, the Quantified Age. The stance of King Canute is not very helpful or realistic; better to ride the waves, if one has sufficient finesse, than to strike attitudes of humanistic defiance and end, in Toynbee's phrase, in the dustbin of history (p. 391).

He found three dangers in the development, however. The first was a confusion of ends and means. Quantifiers wanted to quantify everything (after Lord Kelvin − 'when you cannot express it in numbers, your knowledge is of a meagre and unsatisfactory kind': Spate, 1960b), but some things, like the positions of Madrid and Barcelona in Spanish thought, cannot be treated in that way. Secondly, there was the dogged analysis of trivia, producing platitudinous findings, a fault which Spate recognized is part of all academia, and especially its revolutions: 'Quantified or not, the trivial we will always have with us' (Spate, 1960a, p. 389), and the problem is usually the extreme positions taken up by the protagonists − Robinson's (1961) perks (the hyperquantifiers) and pokes (the hypoquantifiers). Finally, there was the vaunting ambition of the quantifiers, the belief that solution of the world's problems lay just around the corner.

Spate was not the only critic, and he was more generous than many. Burton (1963) identified five types of critic:

1 those who felt geography was being led in the wrong direction;
2 those who felt geographers should stick to their proven tool − the map;
3 those who felt that quantification was suitable for certain tasks only;
4 those who felt that means were being elevated over ends, and there was too much research on methods for methods' sake; and

5   those who objected not to quantification but to the quantifiers' attitudes. He believed, however, that quantification had been proven to be more than a fad or fashion and that geography would develop out of a stage of testing relatively trivial hypotheses with its new tools so that 'The development of theoretical, model-building geography is likely to be the major consequence of the quantitative revolution' (p. 156).

More critical to many geographers than quantification was this issue of theory, and in particular the question of the role of laws in geography. For some, this continued the debate over environmental determinism, which was still active in Britain (Clark, 1950; Martin, 1951; Montefiore and Williams, 1955; Jones, 1956). Jones, for example, carried this debate over to the topic of scientific determinism, and its implications with regard to human free will. Martin (1951) had argued that possibilism is 'not merely wrong but is mischievous' (p. 6) because all human actions are determined in some way, so that in human geography:

> Unless we can assume the existence of laws or necessary conditions similar in stringency to those of physical science, there can be no human geography nor social sciences worth the name, but only a series of unexplainable statements of bare events . . . such laws cannot differ, except in respect of . . . far greater complication, from those of physical science (pp. 9–10).

Jones (1956) pointed out the impossibility of discovering universal laws about human behaviour and indicated the existence of two types of law in physics: the determinate laws of classical physics, which apply macroscopically; and the probabilistic quantum laws which refer to the behaviour of individual particles. Use of the latter would allow for the exercise of free will within prescribed constraints, and would at least allow answers to be offered to the question 'how?' if not to 'why?'. But the question of causality clearly worried many, as indicated by Lewis's (1965) counterargument that 'it is erroneously assumed that causes compel their effects in some way in which effects do not compel their causes' (p. 26).

Golledge and Amedeo (1968) addressed this same problem, by pointing out that critics of law-seeking in human geography applied a definition of a law as a universal postulate which brooked of no exception. They indicated that science recognizes several types of law, and also that the veracity of a law-like statement can never be finally proven, since it cannot be tested against all instances, at all times and in all places. They indicated four types of law which have relevance for human geographers. *Cross-sectional laws* describe functional relationships (as between two maps) but show no causal connection, although they may suggest one. *Equilibrium laws* state what will be observed if certain criteria are met, whereas *dynamic laws* incorporate notions of change, with the alteration in one variable being followed by (perhaps causing) an alteration in another. Dynamic laws may be *historical*, showing that B would have been preceded by A and followed by C, or *developmental*, in which B would be followed by C, D, E etc. Finally, there are *statistical laws* which are probability statements of B happening, given that A exists: all laws of the other three categories may be either deterministic or statistical, with the latter almost certainly the case with phenomena studied by geographers.

None of the papers just discussed was part of an ongoing debate on quantification and theory-building; they appear to have been reactions to attitudes rather than to published critiques (in Britain there were none for several years: Taylor, 1976). There was one debate, however, in the American literature. It was initiated by Lukermann (1958), who was reacting to the views of Warntz on macrogeography (see above, p. 67) and to a paper by Ballabon (1957). The latter had claimed that economic geography was lacking general principles and was 'short on theory and long on facts' (p. 218). McCarty had shown how to conduct research, but stress was placed on the need to study location theory being developed by economists as a source for hypotheses. Lukermann's response was that the main problem in the proposals of Ballabon and Warntz lay in the assumptions behind their hypotheses (Warntz's analogies with physics and Ballabon's with economics) which did not conform to his view of geography as an empirical science. Statistical regularities and isomorphisms with other subject matter do not provide explanations, so that hypotheses derived from such models test only the models themselves (see also Moss, 1970): 'the hypotheses to be tested are neither statistically nor rationally derived; that is, they are derived neither from empirical observation nor from deductions of previous knowledge in the social, economic or geographic fields' (p. 9). Research in economic geography, Lukermann argued, should start with the recording of data on maps.

Lukermann was countered by Berry (1959b) who argued that models, for all their simplifications and unreal assumptions, can offer insights towards understanding the real world: 'A theory or model, when tested and validated, provides a miniature of reality and therefore a key to many descriptions. There is a single master-key instead of the loaded key ring' (p. 12). Lukermann (1960a) was not convinced that models based on assumptions of perfect knowledge and competition, for example, could help in understanding if they were not empirically derived: 'the crucial problem is the construction of hypotheses from the empirical realities of economic geography . . . more light is shed and less truth is sophisticated through inventory than through hunches' (p. 2). King (1960) then entered the debate, pointing out that all laws are really only hypotheses, and that deviations of observed from expected in their testing indicates where the assumptions are invalid. Lukermann responded three times. In the first paper, he showed the lack of consensus in 'explanations' of the geography of cement production in the United States (Lukermann, 1960b) because economic analyses ignored 'Historical inertia, geographical momentum, and the human condition' (p. 5). His second paper (1961), in response to King, presaged some of the arguments developed later by Sack, who worked with him at Minnesota (see below, p. 107), and pointed out that much of the theory being introduced to economic geography (such as Lösch's) was not based on providing understanding of, and explanation for, reality. Finally, he presented a longer paper (Lukermann, 1965) which discussed several aspects of the debate, concluding with the statement that

> Thus, we see scientific explanation as far removed from the context within which the macroscopic geographers would have us put it – the end product of geographic research. Science does not explain reality, it explains the consequences of its hypotheses (p. 194),

and a further call for explanations in geography to be based on observations of reality and not the import of analogies which cannot offer explanations, but only unreal assumptions. Lukermann's basic point, never fully tackled by his critics, was that tests showing conformity between empirical reality and a model were tests of the model only, and could not indicate how empirical reality was created.

The clear difference of opinion between Lukermann and his antagonists over the way in which geographers should seek explanations (which was not about the scientific method itself, but about the inputs to the images of the real-world structure – Figure 3.1) suggests the sort of generation gap discussed in Chapter 1. It is doubtful whether papers such as those of Jones, of Lewis, and of Golledge and Amedeo quieted the fears of those unconvinced by the arguments of the 'quantifiers', any more than Berry and King convinced Lukermann. But the differences soon became a non-issue, at least in the published papers resulting from the research activities of geographers in many topical specialisms. As Burton claimed, by the mid-1960s the changes seen to have been widely accepted, and the regional paradigm had certainly been ousted from its prime position in the publications of human geographers, Increasingly, quantitative and theoretical material using the scientific method came to dominate not only the more obvious journals, such as *Economic Geography* and *Geographical Analysis* (a 'journal of theoretical geography' founded in 1969), but also the prestigious general journals, notably the *Annals of the Association of American Geographers*. (The *Geographical Review* was an early partial 'convert' through the American Geographical Society's sponsorship of the macrogeographers although Berry states that it rejected his early papers with Garrison as 'not geography': Halvorson and Stave, 1978.) Most of the work contributed little to theory, however. It was quantitative testing of theory or model-derived hypotheses in some cases, but with little indication of how good the results were (see p. 72). In others cases, it was quantitative description that could inform theory and model development, but in many instances was merely a series of 'factual reports'. By the 1970s, textbooks were being published which began with discussions of scientific method and quantification before proceeding to the substantive content of the 'empirical science' (Abler, Adams and Gould, 1971; Amedeo and Golledge, 1975).

Spread of the scientific method

Within human geography in the United States, the initial development of systematic studies using the positivist scientific method was very largely focused on economic geography and the associated economic aspects of urban geography. This undoubtedly reflects the relative sophistication of economics within the social sciences, providing a model for geographers to copy, not only to advance their discipline but also to promote its cause in the search for utility to the world of business and government. The long tradition of empirical work in human geography meant, however, that with few exceptions research in the systematic areas mainly comprised the statistical testing of relatively simple hypotheses, with little mathematical modelling or writing of formal theory.

Contemporaneous with these developments in human geography, and an important stimulus for them, was the emergence of a new discipline in the United States – regional science. This was very much the product of one iconoclast scholar – Walter Isard – an economist who built spatial components into his models, in part to provide a stronger theoretical basis for urban and regional planning than had existed previously. In general terms, regional science is economics with a spatial emphasis, as illustrated by Isard's (1956a, 1960) two early texts, and many members of the Regional Science Association are practising geographers. To some, regional science and economic geography are hard to distinguish: the former can be separately characterized by its greater focus on mathematical modelling and economic theorizing, however, whereas geographical work has remained more empirical and less dependent on formal languages. (Initially Isard, 1956b, saw geographers as doing the empirical tests of the regional scientists' models.) Over time, the interests of regional scientists have broadened (Isard, 1975), but the strong theoretical base has remained.

The emphasis on statistical methods in so much of the new work in American human geography led to its partial rapprochement with physical geography. (One of the leading 'quantitative geographers' of the 1960s, Leslie Curry, was trained as a climatologist.) More physical papers were published in the leading journals, more physical geographers were appointed to university departments, geologists such as Krumbein, Leopold, Schumm and Wolman were major sources of quantitative ideas, and there was a common interest in the training of graduate students (LaValle *et al.*, 1967). This common interest in procedures was illustrated by the papers given at a conference held in 1960 on Quantitative Geography, which led to the publication of two volumes (Garrison and Marble, 1967a, 1967b) on the development of methods, one for human geography and the other for physical geography. In the former, for example, Berry introduced the family of factor analysis methods as a way of collapsing and ordering large data matrices; Dacey investigated line patterns and Beckmann the optimal location of routes; Robinson continued his work on the statistical comparison of maps; Mayfield and Thomas extended the analysis of central

place patterns; Marble, Morrill, and Nystuen looked at patterns of movement; and Warntz continued the work on macrogeography. Other conferences and summer schools to train geographers in quantitative techniques
were held at this time (on their impact, see Gould, 1969; Taaffe, 1979), and
American geographers were to the forefront in launching an International
Geographical Union Commission on Quantitative Methods which held
several international conferences and published volumes of papers extending the statistical procedures available to geographers.

Expansion within American geography

The launching of Burton's 'quantitative and theoretical revolutions' took
place in a few topical specialisms within American human geography, so one
of the first tasks for the established 'revolutionaries' was to spread their 'new
Theology' wider through the discipline, by convincing others of the benefits
which quantification and the scientific method could bring to their specialisms. One of the major pieces of advocacy was the NAS/NRC report (1965)
on *The Science of Geography* which was prepared in order to chart research
priorities within the discipline. The case was presented for more 'theoreti-
cal-deductive' work to balance the earlier emphasis on 'empirical-inductive
analysis', the detailed argument being based on four premises:

> (a) Scientific progress and social progress are closely correlated, if not equated.
> (b) Full understanding of the world-wide system comprising man and his natural
> environment is one of the four or five great overriding problems in all science. (c)
> The social need for knowledge of space relations of man and natural environment
> rises, not declines, as the world becomes more settled and more complex, and
> may reach a crisis stage in the near future. Last, (d) progress in any branch of
> science concerns all branches, because science as a whole is epigenetic.
>
>   The social need for knowledge of space relations means an imminent practical
> need. As the population density rises and the land-use intensity increases, the
> need for efficient management of space will become even more urgent (p. 10).

And since, to the members of the committee (E. A. Ackerman, B. J. L.
Berry, R. A. Bryson, S. B. Cohen, E. J. Taaffe, W. L. Thomas Jr, and M. G.
Wolman), geography 'involves the study of spatial distributions on the
earth's surface' (p. 8) then it followed that 'Geographic studies will be irre-
placeable components of the scientific support for efficient space management' (p. 10). The scientific method was being sold to geographers and
geography was being sold to the scientific establishment, which could
provide a great volume of financial research support.

   The committee chose four problem areas within geography to illustrate
the subject's potentials as a 'useful science'. The first of these was physical
geography. The second was cultural geography, which studies 'differences
from place to place in the ways of life of human communities and their
creation of man-made or modified features' (p.23). A major focus in this
area, it was noted, was on landscape development and the diffusion over
space and time of specific cultural features: 'applying modern techniques to

studying the nature and rate of diffusion of key cultural elements and establishing the evolving spatial patterns of culture complexes' (p. 24) was seen as
an avenue for development. The third problem area was political geography, and the committee proposed work on boundaries and resource management. Finally, the committee recognized location theory studies, an
amalgam of work in economic, urban and transport geography in which the
'dialogue' between the empirical and the theoretical had gone furthest,
'revealing the potential power of a balanced approach when applied to other
geographical problem areas' (p. 44). Location theory indicated the nature of
the normal science of the new paradigm, with work on spatial patterns, the
links and flows between places in such patterns, the dynamics of the
patterns, and the preparation of alternative patterns through model-building
exercises which identify efficient solutions.

In the development of the science of geography,

> A major opportunity seen by workers in the location theory problem area is that
> of integrating their work more closely with other geographers as they begin to
> deal with spatial systems of political, cultural, and physical phenomena. . . .
> This could be achieved . . . by the accelerated diffusion of techniques and
> concepts to other geographers, and communication on the definition of research
> problems. The result would be to hasten the confrontation of empirical-
> inductive studies by theoretical-deductive approaches throughout geogra-
> phy. . . . Testing the theory in a variety of empirical contexts should aid in the
> overall development and refinement of viable theories. It should also serve to
> connect geographic progress to local problems more rapidly and more effectively
> (pp. 50–1).

The deductive-theoretical scientific methodology was central to their blueprint for the advancement of geographical research, therefore. All geographers would have a role to play in this movement forward, for:

> Geographers have one other asset that should be capitalized on. Those who have
> been interested in the study of a specific part of the earth (regional geography)
> develop competences for interpreting the physical-cultural complexes of the
> regions that they study. Students of the way a particular part of the earth has
> evolved (historical geography) have other competences for interpreting the
> historical development and modification of a region. These two groups have
> students that are particularly qualified to undertake the field observation and
> field study of problems recognized in a more systematic way and to conduct field
> tests of generalizations arrived at through systematic study. . . .
>   The regional or historical geography specialist who has mastered the technique
> of field observation and historical study thoroughly . . . can make himself indis-
> pensable if he understands the direction in which the generalizing clusters are
> headed and relates his work closely to their growing edges (p. 61).

A clear division of labour was being suggested, comprising theoretical-
deductive 'thinkers' and empirical-inductive 'workers', a division which was

apparently unequal in status and was resented by some as such (James, 1965; Thoman, 1965).

A somewhat similar report, prepared for the Committee on Science and Public Policy of the National Academy of Sciences and the Problems and Policy Committee of the Social Science Research Council, was published five years later (Taaffe, 1970). Also the product of a committee (E. J. Taaffe, I. Burton, N. Ginsburg, P. R. Gould, F. Lukermann, P. L. Wagner), this stressed human geography as 'the study of spatial organization expressed as patterns and processes' (pp. 5–6), incorporating man-environment relationships and cultural landscapes and stressing relevance to planning and other policy issues. Much of the report is constructed to illustrate the nature of human geography. The case

> that there are many opportunities for the expansion and improvement of geographic research. If geography is to have a strong and beneficial impact on the constantly changing patterns of spatial organization of American society, it will be necessary to continue this development (p. 131)

led to six conclusions: (1) greater collaboration among the social sciences; (2) alleviation of geography's manpower shortage; (3) establishment of centres for cartographic training and research; (4) development of remote sensing and related data bases; (5) greater support for foreign area study; and (6) programmes established 'to strengthen the mathematical training of geographers'. Human geography was presented as a relevant component of the social sciences, increasingly sophisticated in its analytical tools, focusing on spatial organization, and offering particular skills in mapping and data acquisition.

One of the systematic areas of geography that was colonized early by the new methods was that part of urban geography which dealt with the internal spatial structure of cities. Until the 1960s, little work had been done on this topic except with regard to commercial land uses (particularly the Central Business District and the relationship of the pattern of suburban shopping centres to the postulates of central place theory): almost no attention was paid to the human content of residential areas, perhaps because geography was seen to be the science of places, not of men. Recognition that 'people live in cities' (Johnston, 1969) led to a growth of interest in residential areas, which gained much stimulus from the work of the urban ecology school of sociologists at Chicago (some of their works had been introduced to geographers earlier – Harris and Ullman, 1945; Dickinson, 1947 – but with little success). The methods of social area analysis gained in popularity, using the techniques being applied elsewhere (Berry, 1964a): a new form of urban geography was initiated which adopted its norms from the functionalist schools of sociology. Society is made up of socio-economic classes, the nature and composition of which are widely accepted, and these classes come to consensual agreements about the allocation of land among competing groups (Johnston, 1971). This type of urban geography in effect became

Figure 3.2 The geographic matrix: each cell – ij – contains a 'geographic fact' – the value assumed by characteristic i at place j. Source: Berry (1964b, p. 6).

a separate systematic branch of the discipline; few people did research in it as well as in other aspects of urban geography.

The argument being advanced by those who sought to spread the new methodology more widely was focused on a common set of procedures to tackle geographical problems. This was described by Berry (1964b), who argued that the geographer's viewpoint emphasizes space, with regard to distributions, integration, interactions, organization and processes. All of its data can be categorized by a single matrix (Figure 3.2) in which places form the rows and characteristics the columns: each cell defines a geographic fact. Berry then suggested that five different types of geographical study can be recognized by focusing on different elements of this matrix: study of a single row (a place) or column (a characteristic); comparison of two or more rows (places) or columns (characteristics); or study of rows and columns together. Adding further matrices, one for each time-period (Figure 3.3.) allows for five further types of study, based on the earlier five but concentrating on changes over time. Thus, he concluded, systematic and regional geography are part of the same enterprise – a repetition of Hartshorne's (1969) arguments – with neither sufficient in itself; the revolution was not changing everything.

Berry's matrices referred only to the characteristis of places, and further matrices could be introduced (Figure 3.4) which show the flows between places, with one matrix for each flow category in each time-period (Clark, Davies and Johnston, 1974). Berry used this extension himself, though he did not formalize it, in his attempt to fuse the procedures for formal and

Figure 3.3 A third dimension to the geographic matrix. Source: Berry (1964b, p. 7)

Figure 3.4 Geographic flow matrices, one per commodity, for each time period. Source: after Clark, Davies and Johnston (1974).

functional regionalization (see above, p. 43) to produce a general theory of spatial behaviour – Berry's (1968) field theory, which he applied in a large study of the spatial organization of India (Berry, 1966). The techniques which he used in this work became widely used in the 1960s, as access to high-speed computers for the handling of very large data matrices became very easy, if not universal, for university academics. They were given the umbrella term of factorial ecology (Berry, 1971), and were widely applied in many aspects of geography (e.g. Zelinsky, 1974), resulting in a methodological unity which was previously unknown across the various systematic specialisms.

The developments of the 1950s in the United States spread into several of the discipline's topical specialisms, and by the mid-1960s use of statistical methods to test hypotheses was common. It was indeed the methods which gave unity to these specialisms, which in terms of substance remained very separate, as identifiable branches witin the geographical enterprise. The decline of interest in work which aimed to integrate the findings of the specialisms into regional syntheses, despite the efforts of Berry and a few others (see also Taaffe, 1974), meant that human geography experienced a centrifugal trend with regard to substance contemporaneously with a centripetal one with regard to procedures. Since the scientific method, and the statistical techniques, were used much more widely than in human geography alone, it seems that the former of these trends was probably the most important.

Transatlantic translation

By the early 1960s the quantitative and theoretical revolutions were having considerable impact beyond the United States, as the result of two agencies. The first was the publication of the work of the American iconoclasts in the major journals. American journals are widely read in Britain (much more than are British journals in America), and emulation was likely. Secondly, and probably more importantly, during the 1950s and 1960s a number of British geographers went to the United States, either as postgraduate students or as visiting staff members. Some returned with the new ideas, which they disseminated among their students and, via the Study Group in Quantitative Methods of the Institute of British Geographers, their fellow academics. (Others stayed in North America: Brian Berry was one.) There was also a local base, mainly in physical geography and arising out of the early use of statistics by climatologists (e.g. Crowe, 1936): as a result, and perhaps somewhat surprisingly, the first undergraduate text in statistics for geographers was written by an English academic (Gregory, 1963). Note that the positivist methodology is entirely implicit in Gregory's book. His Preface speaks only of the geographer's raw material 'becoming progressively more of a quantitative nature' and 'the need to present both data and conclusions in sound quantitative terms' – pp. xiii-xiv. The use of statistical techniques is thus required, but their use in hypothesis testing is not spelled out – see p. 140). In addition, there was some interest in location theory – at University College, London, in the early 1950s, for example (Halvorson and Stave, 1978), and in work on both settlement patterns (Dickinson, 1933; Smailes, 1946) and industrial location (Smith, 1949; Rawstron, 1958).

Although statistics courses were introduced in several university departments of geography in the mid-1960s, and aspects of the scientific methodology were taught in several places by at least a few staff members (Whitehand, 1970), the main focus for the introduction of the 'new geography' to Britain during the early years of the decade was the University of Cambridge. The leaders were R. J. Chorley (a geomorphologist, who had spent some time studying in the United States) and P. Haggett (a Cambridge-trained human geographer, although his early published work was in biogeography, who had also visited the United States and experienced the development there: Haggett, 1965c, p. vi). Their impact on British geography was considerable, through both their teaching and their publications (Gregory, S. 1976). They worked, for example, on the adaptation of certain statistical techniques to geographiccal (both physical and human) ends (Chorley and Haggett, 1965a; Haggett, 1964: Haggett and Chorley, 1969), but their most lasting contribution was probably in editing two collections of papers which resulted from courses that they directed, aimed at introducing the 'new geography' to teachers.

The first of these books – *Frontiers in Geographical Teaching* (Chorley and Haggett, 1965b) – was based on a course given in 1963 which was designed 'to bring teachers and like persons into the University, there to encounter and discuss recent developments and advances in their subjects' (p. xi). In it Wrigley (1965), for example, discussed the changing philosophy of geography and saw the increasing use of statistical techniques as the contemporary development 'of singular importance' (p. 15). He pointed out that techniques of themselves do not form a methodology and that 'Geography writing and research work has in recent years lacked any general accepted, overall view of the subject even though techniques have proliferated' (p. 17). He offered no outline of such a view, however, arguing that eclecticism in mode of analysis was likely to be most productive and that 'the best sign of health is the production of good research work rather than the manufacture of general methodologies' (p. 17). Many of the other chapters interpret geography as if the 'revolutions' had not occurred in the United States: Smith's (1965) on historical geography, for example, is an excellent British companion to the American statement published a decade earlier (Clark, 1954).

Elsewhere in the book, Pahl (1965) introduced the models of the Chicago school of urban sociologists and suggested a social geography in which the prime factor is distance (p. 95), but it was only the chapters by Haggett and by Timms which introduced much of the transatlantic turmoil. Haggett

(1965a) wrote on the use of models in economic geography, both those based on simple views of the world, such as developments on von Thunen's (Chisholm, 1962), and those derived from observations of particular cases (e.g. Taaffe, Morrill, and Gould, 1963). He noted that

> Perhaps the biggest barrier that model builders in economic geography will have to face in the immediate future is an emotional one. It is difficult to accept without some justifiable scepticism that the complexities of a mobile, infinitely variable landscape system will ever be reduced to the most sophisticated model, but still more difficult to accept that as individuals we suffer the indignity of following mathematical patterns in our behaviour (p. 109),

and as a consequence introduced the notion of indeterminacy at the individual level and showed how random variables must be introduced to operational models. His chapter on scale problems (Haggett, 1965b) illustrated methods of sampling and of map generalization from samples. Timms (1965) demonstrated the use of certain statistical techniques for the analysis of social patterns within cities (based on Shevky and Bell's social area analysis, and thus developed independently of Berry's work on this topic – see above, p. 81), pointing out that

> The sciences concerned with the study of social variation have as yet produced few models which can stand comparison with the observed patterns or which can be used to predict those patterns. . . . Prediction rests on accurate knowledge of the degree and direction of the interrelationships between phenomena. This can only be attained by the use of techniques of description and analysis which are amenable to statistical comparison and manipulation. If the goal of geographical studies be accepted as the formulation of laws of areal arrangement and of prediction based on those laws, then it is inevitable that their techniques must become considerably more objective and more quantitative than heretofore (p. 262).

If the majority of the contributors to *Frontiers in Geographical Teaching* were not as committed to the 'new geography' as was Timms (later to become a professor of sociology, as was Pahl), this cannot be said of the editors, who used their epilogue (Haggett and Chorley, 1965) to present a strong case for the 'theoretical revolution':

> We cannot but recognize the importance of the construction of theoretical models, wherein aspects of 'geographic reality' are presented together in some organic structural relationship, the juxtaposition of which leads one to comprehend, at least, more than might appear from the information presented piecemeal and, at most, to apprehend general principles which may have much wider application than merely to the information from which they were derived. Geographical teaching has been remarkably barren of such models. . . . This reticence stems largely, one suspects, from a misconception of the nature of model thinking. . . . Models are subjective frameworks . . . like discardable cartons, very important and productive receptacles for advantageously presenting selected aspects of reality (pp. 360–1).

This view was the dominant one in their next, and substantially more influential, volume (Chorley and Haggett, 1967).

Models in Geography presented a synthesis of most of the work completed before the mid-1960s by adherents to the 'quantitative and theoretical revolutions'. Individual authors had been asked 'to discuss the role of model-building within their own special fields of research' (Haggett and Chorley, 1967, p. 19), which resulted in a series of substantive review essays, some dealing with particular topical specialisms (urban geography and settlement location; industrial location; agricultural activity – there were similar reviews for physical geography), some with particular themes ranging across several specialisms (economic development; regions; maps; organisms and ecosystems; the evolution of spatial patterns), and some with methods and approaches (demographic models; sociological models; network models). A catholic use of the term model was allowed, allowing it as a synonym for a theory, a law, a hypothesis, or any other form of structured idea (see Moss, 1970). The approach was strongly nomothetic, however: as Harvey (1967a, p. 551) expressed it,

> the student of history and geography is faced with two alternatives. He can either bury his head, ostrich-like, in the sand grains of an idiographic human history, conducted over unique geographic space, scowl upon broad generalization, and produce a masterly descriptive thesis on what happened when, where. Or he can become a scientist and attempt, by the normal procedures of scientific investigation, to verify, reject, or modify, the stimulating and exciting ideas which his predecessors presented him with.

All the contributors to the book had clearly chosen the latter course: their focus is on models – on generalization of reality – and mehtods are very much secondary.

The orientation of this significant volume is given by the editors' introduction. (The significance lay in its two uses: first, as a synthesis and argument, the volume was widely read and used by researchers and teachers as a guide; secondly, as a series of major reviews, when republished as a series of paperback volumes, the book was extensively employed as an undergraduate text.) Haggett and Chorley (1967, p. 24) presented the model as:

> a bridge between the observational and theoretical levels . . . concerned with simplification, reduction, concretization, experimentation, action, extension, globalization, theory formation and explanation (p. 24).

It can be descriptive or normative, static or dynamic, experimental or theoretical (see also Chorley, 1964). It forms the basis for a paradigm, which made no attempt 'to alter the basic Hartshorne definition of Geography's prime task' (p. 38) but offered much greater progress:

> the new paradigm . . . is based on faith in the new rather than its proven ability. . . . There is good reason to think that those subjects which have modelled their forms on mathematics and physics . . . have climbed considerably more rapidly than those which have attempted to build internal or idiographic structures (p. 38).

Models in Geography stands as a statement of that faith, and as a major illustration of the expanding use of scientific methods in the systematic fields of human geography.

Although the editors and contributors to *Models in Geography* represent most of the early active participants in the move to change British geography towards a 'more scientific' approach, there were others who are not represented, directly, in that book. Notable amongst them was a group who graduated at the University of Cambridge in the 1950s (Haggett was among them), having been tutored by A. A. L. Caesar (Chisholm and Manners, 1973, p. xi, credit his role, one which continued until the end of the 1970s with a steady stream of productive research workers from St Catherine's College, Cambridge). The group included Michael Chisholm, Peter Hall and Gerald Manners. Only Haggett of this group worked within the 'quantitative revolution'. Chisholm, for example, worked with theoretical developments within systematic fields (e.g. Chisholm, 1962, 1966, 1971a) but with relatively little concern for detailed quantitative analysis (though see Chisholm and O'Sullivan, 1973). Both Hall and Manners were more concerned, as was Caesar it seems, with geographical analysis of contemporary issues, although in some cases the analyses led to attempts at theory-derivation (Hall, 1981a).

Chorley and Haggett's editing, and their joint work on technical developments – such as trend surface analysis (Chorley and Haggett, 1965a) and network analysis (Haggett and Chorley, 1969) – represented a belief in the unity of physical and human geography. This was founded in an assumption that a shared interest in methods and techniques could unite the two – Haggett (1967, p. 664) writes with

> the basic proposition that a wide range of different geographical networks may be usefully analysed in terms of their common geometrical characteristics.

As in North America, while the focus of geographical analysis remained geometry, then physical and human geographers could find common cause (Woldenberg and Berry, 1967). Explanation of the geometry required the study of very different processes, however.

The relatively untouched

The NAS/NRC (1965) report (see above, p. 79) indicated that the two main systematic specialisms within human geography relatively untouched by the developments were cultural and historical geography: in addition, despite Berry's attempt to reframe it (Berry, 1964b), regional geography remained largely separate from the changes in methodological emphasis. Not all cultural, historical and regional geographers ignored the changes occurring elsewhere, of course: some, indeed, were in the vanguard of the 'revolution' – two of the chapters in *Models in Geography*, for example, were written by individuals (David Grigg and David Harvey) who showed a clear focus on historical geography in their earlier research (and later, in the case of Grigg). But in general terms the NAS/NRC report was undoubtedly correct; there is not much evidence of success in winning cultural, historical and regional geographers over to the new methodolgy.

Of the three groups listed in the previous paragraph, the historical geographers have probably been most concerned about their apparent isolation within the discipline. This concern has been summarized by Baker (1972) in terms of the approaches which historical geographers need to consider in greater detail:

> An assumption is necessary here: that methodologically the main advances can be expected from an increased awareness of developments in other disciplines, from a greater use of statistical methods, from the development, application and testing of theory, and from exploitation of behavioural approaches and sources. . . .
>
> Rethinking becomes necessary because orthodox doctrines have ceased to carry conviction. As far as historical geography is concerned, this involves a questioning of the adequacy of its traditional methods and techniques (p. 13).

All of these would have to be followed with care, and the potentials of the methodological developments assessed with caution, but Baker clearly believed there was considerable scope for change, as had already been shown in economic and in social history, and perhaps even more so in archaeology (see Renfrew, 1981). Particular areas of historical geography, perhaps those relating to urban settlements (e.g. Ward, 1971; see also Johnston and Herbert, 1978, p. 20), are perhaps more open to such changes than are others, if for no other reason than the better quality, as well as quantity, of available data, and there is evidence of more work of this type during the 1970s (e.g. Whitehand and Patten, 1977; Johnson and Pooley, 1982). There are possible implications in such work, however: as Baker notes,

> Studies in, for example, 'historical agricultural geography', 'historical urban geography' and 'historical economic geography' seem to offer possibilities of fundamental development, particularly in terms of a better understanding of the processes by which geographical change through time may take place. Such an organization of the subject would view historical geography as a means towards an end rather than as an end it itself (p. 28).

In the early discussions of the relationships between historical geography and the 'theoretical and quantitative revolutions', most attention focused on quantification rather than on theory. As Vance (1978) has pointed out, however, the development of theory does not have to involue 'quantitative abstraction', as work by himself (Vance, 1970), Pred (1977b), Conzen (1971) and others has shown. If available, data can be manipulated to test theories regarding past spatial patterns (e.g. Goheen, 1970), but, as Radford (1981, p. 257) makes clear, theory is the more important:

> In the cities of the nineteenth-century United States, a set of principles . . . was taken to something approaching a logical conclusion.

Such a view assumes that theory is possible in historical geography. As illustrated in Chapter 5, some dispute this – for geography as a whole as well as for historical geography. The positivist method implies objectivity, but the geographer in describing a landscape is subjective:

> In describing a landscape, is he not committed by his past training and his past experiences – by his prejudices, if you will? Just as the portrait an artist paints will tell you much about the artist as well as his sitter, so the description of a countryside will tell you a great deal about the writer (Darby, 1962, p. 4).

Thus to Darby geography is both science and art. It

> is a science in the sense that what facts we perceive must be examined, and perhaps measured, with care and accuracy. It is an art in that any presentation (let alone any perception) of those facts must be selective, and so involve choice, and taste, and judgement (p. 6).

Such a position clearly separated those wedded to an implicitly idealist view of historical geography (see p. 147) from those advancing the cause of a more scientific approach, whether or not it was quantitative.

Cultural geographers would appear to have been less concerned about their apparent drift away from the mainstream of geographical activity than were historical geographers, perhaps because of the lack of any parallel developments to those in geography in anthropology, the discipline with which cultural geographers probably have most contact (see Mikesell, 1967, for a general comparison). (This is a generalization, of course. Anthropological work has experienced major paradigmatic threats, if not changes, notably in the structural work of Claude Lévi-Strauss; see Leach, 1974.) The increasing general interest of geographers in diffusion, stimulated by Hägerstrand's (1968) work and extended into other contexts, such as relatively 'underdeveloped' societies, has led to contact between the spatial analysis and cultural analysis schools of thought, however (Clarkson, 1970). Nevertheless, as Mikesell (1978, p. 1) expressed it, 'Stubborn individualism and a seeming indifference to academic fashion are well-known characteristics' of cultural geographers, whose preferences are for: a historical orientation; a focus on man's role in environmental change, on material culture, and on rural areas: links with anthropology; an individualistic perspective; and field work. (These are all illustrated in a book of essays on *Geography as Human Ecology*: Eyre and Jones, 1966.) Similarly, Porter (1978, pp. 30–31) concludes his review of 'geography as human ecology' with:

> in the past 25 years those interested in the mutual relations of people and environment have taken an interesting journey in search of a satisfactory replacement of environmental determinism. Along the way they have been offered, but generally have declined to use, the shiny wares of gravity and . . . [other] models. . . . They have been impressed by, and at times been perhaps a bit envious of, the accomplishments of their colleagues in the analysis of spatial

organization. . . . [But as the] impulse [in human ecology] is cosmographical, holistic and synthetic, it tends to reject the analytical methods of normal science.

Finally, the situation with regard to regional geography has been surveyed by Paterson (1974), whose essay comprises two main sections – 'On the problems of writing regional geography' and 'Is progress possible in regional geography?' Within the first of the major sections he investigates six problems, which include the growing shortage of subordinate materials (micro-regional studies), and the increasing submergence of regional distinctiveness, though

> only a certain amount of innovation is possible if the regional geographer is to perform his appointed task, which is to convey to his reader the essentials of his region; to illuminate the landscape with analytical light. Landforms and climate are common to all terrestrial landscapes, and human activities to most of them: how shall repetition be avoided? (p. 8). So long as contrasts between region and region (remain), and no matter to what they are attributable, there is work for the geographer to do (p. 16).

He does not conclude, despite the constraints, that there is no possibility of progress: regional geography, he argues, can advance with regard to two criteria – content and insight. Reference to Zelinsky's (1973b) *The Cultural Geography of the United States* illustrates the increased range of content currently being introduced (Zelinsky's book contains a strong theoretical element, that of cultural determinison: see Duncan, 1980); discussion of Meinig's work (e.g. Meinig, 1972) shows the ability of regional geographers to produce fresh spatial insights, although Paterson has to conclude that 'Adventurousness is not a quality that most of us associate with regional geography' (p. 9). Thus:

> The way is open for regional studies which are less bound by old formulae; less obliged to tell all about the region; more experimental and, in a proper sense of the word, more imaginative than in the past, and covering a broader range of perceptions, either popular or specialist (p. 23).
>
> Regional Geography['s] . . . goals are general rather than specific; it is not primarily problem-orientated but concerned to provide balanced coverage, and its aims are popular and educational rather than practical or narrowly professional. Such relevance as it possesses it gains by its appeal . . . to the two universal human responses of wonder and concern. . . . One may recall Medawar's assertion that in science we are being progressively relieved of the burden of singular instances, the tyranny of the particular, and in turn assert that there is a frame of mind on which the particular exercises no tyranny, but a strong fascination (p. 21).

(See also Mead's, 1980, p. 297, strong defence of geographers who 'adopt other lands, . . . share other cultures, . . . [and] make a contribution to the store of knowledge about them'.)

The implication to be drawn from the preceding paragraphs is that there is a difference between, on the one hand, many historical geographers and, on the other, most cultural and regional geographers with regard to the degree to which they have felt 'left behind' or 'relatively untouched' by the changes that occurred during the 1950s and 1960s in other branches of human geography. Certainly the former seem to have been impelled at least to consider the possibility of making methodological changes whereas the latter have continued to work within their established tradition (see also Mikesell, 1973). Not all historical geographers would agree with Baker's analogy from systems theory that simply 'Historical geography has a long relaxation time' (p. 11), however, and Chapter 5 indicates the degree to which they have mounted an attack against the positivist approach, as an alternative to either 'ignoring it in the hope that it will go away' or 'joining it because it can't be beaten'.

There are two issues in the debate between 'quantitative/theoretical' geographers on the one hand and cultural, historical and regional geographers on the other. The first, and relatively trivial, is whether quantitative analysis is feasible. The second, which is much more important, concerns the role of theory. In the positivist method – as applied by human geographers – quantification is a means to an end, which is the articulation of empirically valid theories via hypothesis-testing. It is not the only valid means to that end. If historical and cultural geographers believe in generalization, then they should find philosophical if not technical common ground with the 'quantifiers'. If not, if like many regional geographers they believe in the uniqueness of people in places, then they will see no validity in trying to use theory to illuminate that uniqueness. The importance of this second issue became clear in the 1970s only (see Chapter 5).

Conclusions

Reaction to the regional paradigm began to take shape in the United States during the mid-1950s. To a large extent, the aims of research in human geography were not debated, and relatively traditional definitions of the field were observed: the main issues concerned means and methods. The innovations of the period involved the strengthening of the systematic and topical geographies, and their release from a largely subservient relationship to regional geography, by attempts to develop laws and theories of spatial patterns, using models of various kinds for illumination, and by applying mathematical and, especially, statistical procedures to facilitate the search for generalizations. Whereas the regionalists saw geography as, at most, law-consuming, those of the new persuasion aimed at the production of their own laws, which could be used to explain particular regional outcomes.

These changed means to the geographic end were rapidly accepted in many branches of human geography, particularly in those topical specialisms dealing with economic aspects of contemporary life. They were soon accepted in the growing field which focuses on contemporary social geography, but were relatively ignored in historical geography and almost completely shunned in cultural and regional investigations, which, it was claimed, still focus on unique characteristics of unique places. They also spread into the corresponding fields of study across the Atlantic (and across the Pacific, too), and within little more than a decade British geographers produced a major review volume containing 816 pages of testimony to the enthusiasm of the innovators and the links which they had established with other sciences. But methods are insufficient to sustain an academic revolution unless they can be applied to a coherent substantive core. It is to the search for such a core that the next chapter turns.

4

The search for a focus

As shown in the previous chapter, the changes in human geography which emanated from several centres in the United States during the 1950s were very much concerned with methods of investigation. The ultimate aim of geographical study – as stated in Hartshorne's revised definition (p. 55) – remained the same: indeed, one could argue, from the definitions offered by Ackerman (1963) and the NAS/NRC commitee (1965–see above, p. 79), that human geographers had become even more ambitious in their hopes of explaining 'the world-wide system comprising man and his natural environment'. But within this general ethos, the proximate aims of geographical investigation were not always clear. Systematic studies were in the ascendant, and the intent was to develop valid laws and theories within the implicit positivist framework, but what the exact content of those laws and theories should be was not immediately apparent.

In furthering their analyses, human geographers increasingly sought a clear identity of their own within the social sciences. (Adoption of the positivist philosophy requires that disciplines be identified by their content not their methods, although use of particular procedures – e.g. maps – might be confined to certain disciplines.) They required their discipline to provide a particular viewpoint and contribution to the overall goal of the group of disciplines with which they made common cause. For them, geography required a new focus as well as a new methodology. This new focus was developed around the spatial variable and the study of spatial systems.

Spatial variables and spatial systems

Geography is a discipline in distance, according to the inaugural lecture given by a Scottish professor (Watson, 1955), and its central theme is the relative location of people and places. The importance of relative location within society, and therefore to geography, is seen by Cox (1976) as a result of alterations in societal structure consequent upon technical change. In primitive societies, the main interactions are between relatively isolated groups of men and their physical environments, so that a natural early focus for geographical work was the relationships between societies and 'a spatially differentiated nature' (p. 192). With technological advancement, however, the main links are between men and other men. Interdependence within and between societies increases as a consequence of the more complex differentiation between places which reflects the division of labour, so that the most important facts in modern human existence relate to spatially differentiated societies, not to a spatially differentiated nature. It is this interdependence between groups living in different places which creates the patterns of human occupance on the surface of the earth, and provides the basic subject matter for human geographers.

The focus on spatial arrangements of spatial structures – the areal differentiation in human activities and the spatial interactions which this produces – and the role of distance as a variable influencing the nature of those arrangements can be identified in the textbooks of the 1960s and 1970s which summarized for the next generation of students the contemporary activities of human geographers. A pioneer among such texts was Haggett's (1965c), whose depiction of pattern and order in spatial structures was phrased within a decomposition of nodal regions into five geometrical elements; a sixth element was added in the second edition of the book (Haggett, Cliff and Frey, 1977).

The geometrical elements in Haggett's schema (Figure 4.1) assume a spatially differentiated society within which there is a desire for interaction; people in place X want to trade with those in place Y, for example, whereas those in place Z want goods and services which they themselves cannot provide. This results in patterns of *movement* – of goods, of people, of money, of ideas, and so on – between places and so the first element in the

Figure 4.1 The elements in Haggett's schema for studying spatial systems: A movement; B channels; C nodes; D hierarchies; E surfaces; and F diffusion. Source: Haggett, Cliff and Frey (1977, p. 7).

analysis of nodal regions involves the representation of the patterns of movement. Some movement is unimpeded – aircraft can move in all directions – but most is channelled along particular route corridors. Thus the second element in the analysis involves characterization of the movement *channels* or networks. Networks comprise edges and vertices; in a transport system, many of the latter are the *nodes*, the organizational nexus. Their spatial arrangement forms the third element in the decomposition of the nodal region, whereas the fourth investigates their organization into hierarchies, which define the importance of places within the framework of settlements. Finally, in the original scheme, there are the *surfaces*, the areas of land within the skeleton of nodes (settlements) and networks (routes) which are occupied by land uses of various types and intensities.

Patterns in the human occupation of the earth's surface change frequently in 'modern' societies, and the spatial order in such changes forms the sixth element in Haggett's revised schema. Change does not take place uniformly over space in most circumstances: usually it originates at one or a few localities from whence it spreads to others, along the movement channels, through the nodes, across the surfaces, and down the hierarchies. The processes of change over space and time thus involve spatial *diffusion*.

Haggett and his associates stress that geography is a science of distributions, and their emphasis is on the regularities in various elements of these distributions. (The first volume of the text was very reliant on work in other disciplines. The second illustrates the volume of work done by geographers in a short time: this work has strongly influenced recent developments in aspects of archaeology (Renfrew, 1981).) Other texts have taken a similar general view of the study of what have been termed spatial systems, although perhaps emphasizing different aspects of those systems. Morrill (1970a), for example, chose a title – *The Spatial Organization of Society* – which clearly emphasizes his view of the role of geographical analysis in the larger task of the social sciences, 'understanding society'. According to him, the core elements of human geography are:

> Space, space relations, and change in space – how physical space is structured, how men relate through space, how man has organized his society in space, and how our conception and use of space change (p. 3).

In this context, space has five qualities relevant to the understanding of human behaviour: (1) distance, the spatial dimension of separation; (2) accessibility; (3) agglomeration; (4) size; and (5) relative location. Put together, these can be used to build theories, such as those on which Garrison based his work (p. 63).

> Virtually all theory of spatial organization assumes that the structure of space is based on the principles of minimizing distance and maximizing the utility of points and areas within the structure, without taking the environment, or variable content of space, into account. Although the differential quality of area is interesting and its effect on location and interaction is great, most of the

observable regularity of structure in space results from the principles of efficiently using territory of uniform character. The theoretical structures for agricultural location, location of urban centres, and the internal patterns of the city are all derived from the principle of minimizing distance on a uniform plane (p. 15).

According to this view, all location decisions and decisions about the use of land are taken in order to minimize the costs of movement. The spatial approach to understanding society assumes a world in which one variable is predominant as an influence on human behaviour – distance – and it seeks to account for observed spatial patterns within this framework. And so in Morrill's book:

> the explanation of spatial structure proceeds from the deductive – what would occur under the simplest conditions – to the inductive – how local factors distort this 'pure' structure. To begin with, all the local variation may introduce is a risk of missing the underlying structure. Most theory of location therefore stresses the spatial factors – above all, distance – which interact to bring about the regular and repetitive patterns (p. 20).

Morrill's book, which by no means presented an isolated, individual view (it has been widely used as a text, and was revised and republished four years after the original edition; a third edition appeared in 1979), thus suggests an organizing framework for geographical scholarship centred on the single variable of distance. His summarizing 'theory' of spatial structures proceeds as follows:

1　Societies operate to achieve two spatial efficiency goals:
  (i)　　to use every piece of land to the greatest profit and utility; and
  (ii)　　to achieve the highest possible interaction at the least possible cost.
2　Pursuit of these goals involves four types of location decisions:
  (i)　　the substitution of land for transport costs when seeking accessibility;
  (ii)　　substituting production costs at sites for transport costs when seeking markets;
  (iii)　　substituting agglomeration benefits for transport costs; and
  (iv)　　substituting self-sufficiency (higher production costs) and trade (higher transport costs).
3　The spatial structures resulting from these decisions include:
  (i)　　spatial land-use gradients; and
  (ii)　　a spatial hierarchy of regions.
  These are somewhat distorted by environmental variations to produce:
  (iii)　　more irregular but predictable patterns of location.
  Whereas over time distortion may result from:
  (iv)　　non-optimal location decisions; and
  (v)　　change through processes of spatial diffusion.

Like Haggett, therefore, Morrill stresses the geometry of man's organization of his activities on the earth's surface, but whereas Haggett emphasizes pattern and geometry Morrill pays more attention to the decision-making processes which would produce the most 'efficient' pattern, as an underlying basis for the imperfect examples of that pattern which are observed in the 'real world'. Other texts (such as Abler, Adams and Gould, 1971, which has a stronger focus on positivist methodology: see also Gould, 1977, 1978) have followed Morrill's lead.

These textbooks illustrate the centrality of space or distance as the major focus of geographical interest in the 1960s. The emphasis on pattern, evident especially in Haggett's book, was noted by King (1969) in a major review of 'The geographer's approach to the analysis of spatial form . . . the mathematics which are used and the geometrical frameworks which are favoured' (p. 574). His focus was descriptive mathematics which represent 'what is' rather than 'what should be'. He realized that 'when they are pursued to their extremes in very formal terms these studies run the risk of appearing as seemingly sterile exercises in pure geometry' (p. 593; see also King, 1976, 1979b) but felt that geographers had not proceeded very far in providing process theories which would account for observed spatial patterns. The school of thought which he was reviewing involved working backwards, finding what order there was to explain rather than deducing what the world should be like from knowledge of human behaviour.

Spatial theory

Just as the methodological developments reviewed in the previous chapter proceeded without any clear guidelines in terms of a programmatic statement, so the growth of the spatial viewpoint similarly lacked any manifesto. (Watson's paper of 1955, mentioned above, was not widely referenced.) The only attempt to provide such a lead – apart from general statements about geography and geometry, such as Bunge's (1962) – was provided by Nystuen (1963), in a paper which was not widely read until its reprinting in 1968.

Nystuen's objective was 'to consider how many independent concepts constitute a basis for the spatial point of view, that is, the geographical point of view' (p. 35 – all page references are to the 1968 reprint) so that rather than look at the 'real world', with its many distorting tendencies, he sought clarity in considering abstract geographies. To illustrate his deduced basic concepts he used the analogy of a mosque completely lacking furniture (i.e. an isotropic plain) in which a teacher chooses a location at random. His students then distribute themselves so that they can see and hear him; their likely arrangement is in semi-circular, staggered rows facing him, with a greater density close to him. This arrangement has three characteristic features:

1 Directional orientation – they all face the teacher, the better to hear his words and to perceive his expressions;

2 Distance – they cluster around him, because the effectiveness of his voice diminishes with distance; and

3 Connectiveness – they arrange themselves in rows, so organized that each person has a direct line of sight to the teacher.

The third of these, connectiveness, is in part a function of distance and direction, but not entirely so. As Nystuen expresses it:

> A map of the United States may be stretched and twisted, but so long as each state remains connected with its neighbors, relative position does not change. Connectiveness is independent of distance and direction – all these properties are needed to establish a complete geographical point of veiw (p. 39).

In addition

> Connections need not be adjacent boundaries or physical links. They may be defined as functional associations. Functional associations of spatially separate elements are best revealed by the exchanges which take place between the elements. The exchanges may often be measured by the flows of people, goods, or communications (p. 39).

In the mosque, therefore, the connectivity between teacher and student involves not only a direct line of sight between them but also a flow from one to the other, in this case of ideas.

These three concepts – direction, distance and connectiveness – are the necessary and sufficient ones for the construction of Nystuen's abstract geography, which is grounded in the study of sites (abstract places) rather than of locations (real places).

> The terms which seem to me to contain the concepts of a geographical point of view are *direction or orientation, distance,* and *connection or relative position.* Operational definitions of these words are the axioms of the spatial point of view. Other words, such as pattern, accessibility, neighbourhood, circulation, etc., are compounds of the basic terms. For abstract models, the existence of these elements and their properties must be specified (p. 41).

Nystuen was unsure whether these three comprised the full set of necessary and sufficient concepts for a geographical argument; boundary, he felt, might be a primitive concept and not a derivative of the basic three (see also Papageorgiou, 1969). But his general case was that arguments in human geography could be based on a small number of such concepts, a case that was implicitly accepted by much of the work in 'the spatial tradition', but which was rarely explicitly referred to.

An alternative approach to the writing of theory in human geography has been taken by Haynes (1975), based on the mathematical procedures of dimensional analysis. Five basic dimensions are defined – mass, length, time, population size, and value – and these are manipulated to indicate the validity of functional relationships, such as distance-decay equations (see p. 102), by checking their internal consistency. This approach is defended by the statement that:

> Although most quantitative geographers would probably claim to be engaged in the discovery of relationships, it appears that geography has not passed the first stage (in the development of a science) with any degree of rigour. . . . With no clear idea of which variables are relevant and which particular characteristics in a

system should be isolated, it is pragmatic to define our measurement scales with regard to a particular set of observations rather than the other way round. The method of physical science . . . is a superior system, as measurements can be interpreted exactly, different results compared, and experiments replicated (p. 66).

This is clearly a deductive approach to research. It is not as closely tied to a 'spatial view' as Nystuen's but has the same end in view – the derivation of a set of fundamental concepts which can form the basis for the writing of geographical theory, to be tested in the 'real world' (see also Haynes, 1977, 1982).

These attempts at isolating human geography's primary concepts differ from most contemporary efforts at producing geographic theory which, according to Harvey (1967b, p. 212), were 'either very poorly formulated or else derivative'. Central place theory, for example, was based entirely on postulates from economics about how people behave as 'economic men', to which the basic geographic concept of distance was added, producing a theory about the size and spacing of settlements. The attraction of central place theory to many was indeed that geographers could apparently contribute their own basic concept to the development of theory, and need not be totally dependent on other disciplines for concepts, as the regional paradigm had perhaps suggested that they should be (see p. 65). For Harvey (1970) geography has a group of such concepts – location, nearness, distance, pattern, morphology; most of them compounds of Nystuen's basic terms – which could form indigenous elements in theory-building, allowing development of integrated social science theories drawing concepts of equal status from all component disciplines.

An example of the development of theory involving both derivative and indigenous concepts is the field of diffusion research, which received considerable attention in the late 1960s (Brown, 1968). The basic behavioural postulate, taken from sociological research findings, was that the most effective form of communication about innovations is by word of mouth. Geographers expanded on this by introducing the effect of distance: most inter-personal communication, it was found, is between neighbours, so that information about innovations should spread outwards in an orderly fashion from the locations of the initial adopters. Pioneer work on this hypothesis had been carried out in Sweden by Hägerstrand, who introduced it to the Washington school in the 1950s, where it was taken up by Morrill (1968): Hägerstrand's own major work was made available in English translation in 1968. Much work has since been done, both on the process of diffusion and, even more, on patterns of spatial spread which, it is assumed, result from diffusion processes (see Abler, Adams and Gould, 1971, Chapter 11). An extended review of this literature in provided by Brown (1981), who has introduced a more behavioural perpective into the processes involved (see p. 139).

Social physics and spatial science

Of the three basic concepts identified by Nystuen, two – distance and connectivity – received most attention from those advocating geography as a spatial science. Direction was relatively ignored, except for some work on migration patterns (Wolpert, 1967) which included a seminal statement by Adams (1969). (Direction is used here in cardinal terms. To some extent, any discussion of movement patterns which identified destinations more precisely than in terms of distance from origins involved directional analysis.)

By far the greatest volume of work on spatial science followed the lines of research established by the social physics school (p. 66). The relationship between distance and a variety of types of interaction – migrations, information flows, movements of goods, etc. – had been identified by several workers in the nineteenth century, such as Carey (1858), Ravenstein (1885; Grigg, 1977), and Spencer (1892). The impact of these early writings is not clear. McKinney (1968), for example, has suggested that Stewart and others were unaware of the seeds of the 'gravity model' and 'population potential' ideas in the works of Carey and Spencer, and claimed that 'current geographers could learn much' (p. 105) from their publications. Warntz (1968) retorted that Stewart was well aware of such writings, although McKinney in a rebuttal pointed out that Stewart referred to them in his 1950s papers, but not in those of the 1940s. Ravenstein's papers, on the other hand, were clearly influential on later generations of researchers into migration patterns. (It is of interest that the pioneering work in distance-decay and the gravity model, as with much of the rest of spatial science, was done outside geography: as Tocalis, 1978, p. 124 expresses it, 'geographers' contributions to the theoretical evolution of the gravity concept were minor', although one major contributor – Alan Wilson – 'became a geographer'.)

Apart from Stewart, whose pioneering efforts were referred to in the previous chapter, the most influential figure on social physics after World War II was probably Zipf, who devised a 'principle of least effort'. According to this, individuals organize their lives so as to minimize the amount of work which they must undertake (Zipf, 1949). Movement involves work, and so the minimization of movement is part of the general principle of least effort. To explain this, he expanded on Stewart's finding that with increasing distance from Princeton, fewer students from each state attended that university. Two aspects of work are involved in going to university: (1) the work involved in acquiring information about the university; and (2) the work involved in actually travelling there. Thus the greater the distance between a person's home and Princeton, the less he is likely to know about the university and the less prepared he will be to travel there. The validity of this distance-decay argument was tested against many other data sets: material on the contents of newspapers and on their circulation, for example, illustrated the expected distance bias in flows of information, and data on movements between places showed that the greater the distance separating them, the smaller the volume of inter-place contact.

Zipf called the regularity which he had identified the P_1P_2/D relationship, and Stewart noted the analogy between this and the Newtonian gravity formula

$$F_{ij} = k \frac{M_i M_j}{d_{ij}^2}$$

where M_i and M_j represent the mass at places i and j respectively; d_{ij} is the distance separating i and j; k is a coefficient of proportionality (a scaling coefficient); and F_{ij} is the gravitational force between places i and j. For interaction patterns, this was rewritten as

$$I_{ij} = k \frac{P_i P_j}{d_{ij}^2}$$

where P_i and P_j are the populations of places i and j respectively, k and d are as in the previous equation, and I_{ij} is the amount of interaction between places i and j. Much work was done which involved fitting this equation to flow data, as shown in the reviews by Carrothers (1956) and Olsson (1965). In order to achieve reasonable statistical fits, the various elements of the equation had to be weighted, in the form

$$I_{ij} = f(P_i^a P_j^b d_{ij}^c)$$

where a, b, and c are weights estimated from the data set being analysed. Because different values of a, b and c were produced in almost all studies, it was claimed that the so-called gravity model of interaction was 'an empirical regularity to which it has not yet been possible to furnish any theoretical explanation' (Olsson, 1965, p. 48: such an explanation, in statistical terms, was offered by Wilson, 1967). In effect, the influence of distance apparently varies from place to place, from population to population, and from context to context, in the use of the model to represent migration flows. The influence, as shown by the established connectivities, would seem to be virtually universal; what no theory has been able to account for is the variability in the strength of its impact.

It was not only in social physics, and the work of Stewart, Zipf and others, that distance was receiving attention as an important variable in the years after the Second World War. As Pooler (1977) has indicated, both econo-mists and sociologists were becoming increasingly aware of its influence on behaviour, the former in the location theories of Weber, Hoover, Lösch and others, which stimulated the work of Garrison and his associates (p. 62), and also of Isard, and the latter in the studies of the Chicago school, out of which developed a large literature on urban residential patterns (Johnston, 1971, 1980b). And so, according to Pooler:

> a number of geographers became aware of the spatial enquiries that were being undertaken in a social science context outside their own discipline and, upon

realizing their relevance to geography, proceeded to emulate them. The appearance of the spatial tradition was prompted, not by discoveries from within geography, but by an awareness and acceptance of investigations external to the discipline. The space-centred scientific enquiries of other social sciences became paradigms for geographers, simply because those enquiries, being spatial, were seen to be of relevance by some practitioners (p. 69)

thereby fitting in with both the philosophical framework outlined by Schaefer and the 'quantitative revolution'. In their adoption of spatial viewpoints pioneered in other disciplines, human geographers were frequently selective in what they imported from the various sources. Thus the work on urban residential patterns concentrated on certain aspects of the urban sociology of the Chicago school, for example. Robert Park's dictum relating social distance to spatial distance stimulated much work on the measurement of residential separation (Peach and Smith, 1981), but the social Darwinism and ecological theory underlying this dictum was largely ignored (though see Robson, 1969; Entrikin, 1980) and the humanistic concerns in Park's work have only recently been identified (Jackson and Smith, 1981).

In their distance-based analyses, various social scientists looked not only at the influence of distance itself but also at the meaning and measurement of that concept. Stouffer (1940), for example, established a relationship in which migration between X and Y was accounted for not so much by the distance between them but rather by the number of intervening opportunities. In effect, he was measuring distance in terms of opportunities; the greater the number of opportunities available locally, the less work that has to be expended in moving to one. Other took up this flexible approach to measurement of the basic variable. Ullman (1956), for example, developed a schema for analysing commodity flows in which the amount of movement between two places was related to three factors:

1 complementarity – the degree to which there is a supply of a commodity at one place and a demand for it at the other;
2 intervening opportunity – the degree to which either the potential destination can obtain similar commodities from a nearer, and presumably cheaper, source or the potential source can sell its commodities to a nearer market; and
3 transferability – for complementarity to be capitalized it is necessary for the movement to be feasible, given channel, time and cost constraints.

This schema was not as easily fitted statistically as the 'gravity model', but its relationship to that is clear (Hay, 1979b). Fitting such models also requires accepting that the influence of distance, as time and cost, varies from place to place and from time to time, as shown by Abler (1971), Forer (1974), and Janelle (1968, 1969).

All of this work on the analysis of various movement patterns was stimulated not only by its obvious relevance to the spatial science movement within geography, and its use in the development of location theories, but also by its clear applicability in forecasting contexts. The planning of land-use patterns and transport systems (especially road systems) became increasingly sophisticated, in a technical sense, during the 1950s and 1960s, first in the United States and then in Britain. Initially, data were collected to show both the traffic-generating power of various land uses and the patterns of interaction between different parts of an urban area, with the gravity model being used to describe the latter. Future land-use configurations were then designed, their traffic-generating potential derived, and the gravity model used to predict the patterns of flows and the needed road systems. Later models, most of them based on one initiated by Ira Lowry, were able to assess different configurations in terms of traffic flows, thereby suggesting the 'best' directions for future urban growth (see Batty, 1978).

The demands for sophisticated planning devices stimulated much research based on the gravity and Lowry models. American economists initiated this, but it was later taken up by British workers, led by, *inter alia*, Alan Wilson, who in 1970 was appointed to a professorship of geography at the University of Leeds, thereby becoming a professional geographer, although he had no training in the discipline. He derived the gravity model mathematically, thereby giving it a stronger theoretical base (Wilson, 1967), and developed the Lowry model into a more general set of models concerned with location, allocation and movement in space (see below, p. 120).

Much of this work involved collaboration between academic geographers and practising planners, and led to developments which paralleled those in regional science in the United States. Two new British journals catered for these and related research areas – *Regional Studies* and *Environment and Planning* – both of which attract contributions and readers from other social sciences. Thus the new geographical methodology was proving of considerable applied value. Careers as planners became extremely popular among geography graduates for a few years, notably in the late 1960s and early 1970s, and the growing desire of academic geographers for their discipline to equal that of other social sciences in its policy-making relevance seemed well on the way to fulfilment.

Spatial science and spatial statistics

The developments just outlined were characterized by a large volume of technical research within human geography, concerned not only with fitting the gravity model but also with a range of problems associated with the description of spatial patterns and connectivities (for example, Haggett and Chorley, 1969). Although Bunge (1962), Harvey (1969a) and others argued that the language for the analysis of spatial form is geometry, much of this work in fact concentrated on the application of descriptive and inferential statistics to spatial problems, in line with the strong empirical tradition in geographical work.

Most researchers, including Garrison (1956a), accepted that statistical procedures developed for other fields of enquiry could be adopted for geographical investigations without difficulty. For some, it was necessary to introduce modifications – as in spatial sampling (Berry and Baker, 1968) – but, despite the early work of Robinson (1956; see p. 62), it was generally assumed that there were no technical problems involved in the application of standard procedures to spatial data sets. A number of student texts appeared, especially in the late 1960s, and these made little or no reference to any peculiarities of geographical data; they differed from the texts produced by other social scientists only in the nature of their examples.

During the late 1960s, however, a group of researchers at the University of Bristol began to question the widespread assumption about the relevance of most statistical procedures in geographical investigations. For some time, statisticians had been aware of difficulties in applying the general linear model – notably in its regression and correlation form – to the time-series data used in economic analysis and forecasting. The major issue was that of autocorrelation. One of the assumptions of the model is that all observations are independent of each other; the magnitude of one reading on a variable should in no way influence that of any other. Such autocorrelation clearly exists in most time series: the value of the retail price index at one date, for example, has a very strong influence on what the value will be at nearby later dates, and is itself influenced by the value at earlier times. (This issue was first made clear to geographers in the discussion in Chisholm, Frey and Haggett, 1971, p. 465.) Because of such interdependence between adjacent observations, conventional regression methods could not be used: autocorrelation led to biased regression coefficients and therefore cast doubt on the validity of any forecasts.

The Bristol group argued that this autocorrelation problem also existed with spatial data sets, and further that it was much less tractable than was so with temporal series. Time progresses in one direction only, but space is two-dimensional, and the independence requirement can be violated in all directions around a single point. Spatial autocorrelation, therefore, can involve all neighbours influencing all others with regard to the values of a particular variable, as recognized by some statisticians (e.g. Geary, 1954) and hinted at by geographers such as Dacey (1968) whose work stimulated the Bristol investigations. The group expended considerable effort over a number of years on this problem (Cliff and Ord, 1973, 1981), and the research was extended over a wide range of spatial applications, including the well-known gravity model (Curry, 1972; Johnston, 1976a; Sheppard, 1980; Fotheringham, 1981).

Recognition of the spatial autocorrelation problem indicated severe constraints on the application of conventional statistical procedures to geographical analyses, since the argument was that application of regression methods, and others based on them such as principal components and factor analyses, was invalid with spatial data sets (Haining, 1980). As Harvey (1969a) put it:

> The choice of the product-moment correlation coefficient for regionalization problems appears singularly inappropriate, since one of the technical requirements of this statistic is independence in the observations. Since the aim of such regionalization is to produce contiguous regions which are internally relatively homogeneous, it seems almost certain that this condition of independence in the observations will be violated (p. 347).

The general tenor of the argument was obvious for, as Gould (1970a) noted, spatial autocorrelation was the order that geographers were seeking to establish with their laws and theories. On realizing the force of the case, some accepted that conventional statistical procedures could not be applied in their work (e.g. Berry, 1973b) and the Bristol group omitted them from their rewrite of Haggett's major text (Haggett, Cliff and Frey, 1977). Most continued to apply the methods, however, either in ignorance of the autocorrelation case or on the grounds that the biased-coefficients problem refers only to the use of the general linear model in forecasting and prediction, and does not affect use of the procedures for description (see Johnston, 1978a) and for certain types of ecological analysis (Johnston, 1982d).

The forecasting and prediction issue was basic to the work of the Bristol group, however, since their activities were focused on what they termed spatial forecasting, which involved development of procedures for estimating how trends – in the spread of a disease, for example, and in prices and unemployment – would proceed through time and over space (Haggett, 1973; Cliff *et al.*, 1975). They organized a major symposium on this (Chisholm, Frey and Haggett, 1971) and have stimulated a considerable volume of work, much of it technical, on the problems of identifying and forecasting spatio-temporal trends (e.g. Bennett, 1978b). Hägerstrand's work on diffusion patterns is the natural base for much of this research, which focuses on the patterns of spread rather than on their generating processes. There is now a large literature on the technial aspects of forecasting (e.g. Bennett, 1979), and on deducing 'Spatial processes' from mapped patterns (Haining, 1981).

The nature of such spatial forecasting has been reviewed by Hay (1978), who challenges what he sees as its fundamental assumption, that phenomena behave coherently in both time and space. He questions the assumed stability of inter-place interrelationships over time present in analyses such as that of Martin and Oeppen (1975) on market-price variations, and argues instead for consideration of the value of catastrophe theory. In this, small changes in the control variable(s) can lead to major changes in the dependent variable being studied (such major changes are the catastrophes). If catastrophes occur, then the linear extrapolations typical of the work of the Bristol group have clear limitations as forecasting procedures; as yet, relatively little work has been done by geographers on the application of this relatively new area of mathematics to their problems (Wilson, 1976a, 1981a).

One final statistical, and substantive, problem in the spatial approach has been identified by Harvey (1973, p. 40); he calls it the problem of confounding. This involves a non-experimental situation in which there are two interdependent variables operating on a third, and it is extremely difficult to establish the relative importance of either. Thus, for example, geographical studies show that city residents tend to marry people who live nearby, whereas sociological investigations show that most marriages are between people of the same social background. Amalgamation of these two findings would suggest that people tend to choose spouses from among those of similar social backgrounds and who live in the same area as themselves. But, as Harvey (1973) expresses it:

> Now it so happens that people of the same class tend to live in proximity to one another. How then are we to distinguish how much the spatial variable contributes, and how much the personality variables contribute? . . . We do not appear to possess adequate non-experimental research designs to allow us to handle this sort of problem in any but the crudest fashion (pp. 40–1).

Does the spatial influence produce the social class effect? Does the social class effect – and the associated spatial segregation of classes – produce the distance effect? Or are both operating, and in what relative strength? The problems of answering these questions raise important doubts, at least in some minds, about the significance of the spatial point of view to the study of human behaviour.

Countering the spatial separatist theme

Opposition to the widespread view of human geography as a spatial science developed largely as a counter to the claims for separate status for such a discipline. Some, such as Crowe (1970), saw the use of the spatial variable in nomothetic studies as a naive spatial determinism which paralleled the earlier environmental determinism. Other based their criticisms on the implicit divisibility of the social sciences in the spatial claims, and it is these criticisms which form the subject of the present section.

The most sustained argument against the view that geography is a spatial science – what he calls the 'spatial separatist' theme – has been presented in a series of papers by Sack, a former associate Lukermann at the University of Minnesota (see p. 76). There are three dimensions of reality – space, time and matter – and geography, according to the spatial separatist view, is the science of the first of these. But according to Sack, space, time and matter cannot be separated analytically in an empirical science which is concerned to provide explanations. Thus in his first paper he set out to show that geometry is not an acceptable language for such a science (Sack, 1972). Geometry is a branch of pure mathematics which is not concerned with empirical facts; its laws are static laws, with no reference to time, and they are not derivable from any dynamic or process laws. Geographic facts have geometric properties (locations) but if, as Schaefer proposed, geographical laws are concerned only with the geometries of facts, then they will provide only incomplete explanations of these facts. (To illustrate this contention, Sack used an analogy of chopping wood. If the answer to the question 'why are you chopping wood?' is 'because the force of the axe on impact splits the wood' then it is a static, geometric law, but if the answer is 'to provide fuel to produce heat' then it is an instance of a process law, which incorporates the geometric law. In this analogy, a process law is equated with the intention behind an action.) The laws of geometry, according to Sack, are sufficient to explain and to predict geometries, so that if geography aimed only to analyse points and lines on maps it could be an independent science using geometry as its language. But 'We do not accept the description of the changes of its shape as an explanation of the growth of a city' (p. 72) so that 'Geometry alone, then, cannot answer geographic questions' (p. 72) leading to the conclusion that:

> To explain requires laws and laws (if they are valid) explain events. Since the definition of an event implies the delimitation of some geometric properties (all events occur in space), the explanation of any event is in principle an explanation of some geometric properties of events (p. 77).

Thus geography is closely allied with geometry in its emphasis on the spatial aspects of events (the instances of laws), but geometry alone is insufficient as a basis for explanation and prediction since no processes are involved in the derivation of geometries.

Bunge (1973a) responded to this statement, claiming that spatial prediction was quite possible with reference to the geometry alone, as instanced by central place theory and Thunian analysis. Such geometries provide 'classic beauty', and 'purging geometry from geography reduces our trade to no apparent gain' (p. 568). Sack's (1973a) reply was that the static laws espoused by Bunge are only special cases of dynamic laws having antecedent and consequent conditions, and that:

> Although the laws of geometry are unequivocally static, purely spatial, non-deducible from dynamic laws, and explain and predict physical geometric properties of events, they do not answer the questions about the geometric properties of events that geographers raise and they do not make statements about process (p. 569).

See also the response by Coffey (1981), who argues for an independent analysis of space.

Sack did not argue, as Bunge supposed, that geometry should be purged from geography, but only that space should not be considered independently from time and matter. He developed this theme (Sack, 1973b) with the contention that:

> for a concept of physical space to conform to the rules of concept formation and be useful in a science of geography every instance of the geometric or spatial terms must be connected or related to one or more instances of non-geometric terms (to be called substance terms) (p. 17).

Thus physical distance is not a concept in itself: it is necessary to know the terrain which a road crosses, for example, in order to assess the significance of its length in a gravity model – geometry alone is not enough (hence the considerations of the meaning of distance referred to above – p. 103 – and the work of Mackay (1958) on the influence of boundaries on movement patterns). Since there is no such thing as empty physical space so there are no frictions of distance per se. There are frictions which demand work in crossing a substance, but it is the substance itself which creates the frictions, and the context in which it is being crossed, not simply the distance: 'There are frictions and there are distances, but there is no friction of distance' (p. 22). Geography, according to Sack, is concerned to explain events and so it requires substantive laws: such laws may contain geometric terms, such as the frictions of crossing a certain substance, but these terms of themselves are insufficient to provide explanations.

The spatial separatist approach proposes an independent position for geography within the social sciences based on its use of geometry, but Sack (1974b) contends that 'The spatial position's aim of prying apart a subject matter from the systematic sciences by arguing for spatial questions and spatial laws does not seem viable' (p. 446). Instead, two types of law relevant to geographical work must be identified (Sack, 1974a). *Congruent substance laws* are independent of location: statements of 'if A then B' are universals which require to spatial referent. *Overlapping substance laws*, on the other hand , involve spatial terms: 'if A then B' in such cases contains some specific reference to location. Both types are relevant and necessary in providing the answers to geographical questions, so no case can be made for a necessary 'spatialness' to the substance laws of human geography. Further, it is necessary to realize that 'Space is an essential framework of all modes of thought' but that 'geographic space is seen and evaluated in different ways at different times and in different cultures' (Sack, 1980, pp. 3–4).

May (1970) has also argued against Schaefer's claim that geography is the study of spatial relations:

> If we extend Schaefer's argument to include time, and assign the study of temporal sequences or relations to the historian, then the only conclusion respecting this matter that can be drawn is that economic, social, political, and other relations must be non-spatial and non-temporal. Hence economics, sociology, political science, etc. are non-spatial, non-temporal sciences. But this is absurd . . . insofar as economics qualifies as a science possessing empirical warranty, then its generalizations must apply to given spatio-temporal situations (p.188).

If, then, all sciences have a spatial content, what is there left with which to define a separate discipline of geography? May lists five possibilities.

1 Geography is a 'super-science' of spatial relations, 'a generalizing science of spatial relations, interactions and distributions' (p. 194) drawing on the findings of other sciences. This would leave the latter truncated and their studies unfinished. In any case such an approach has been unsuccessful; 'the

issue of the conception of geography as a generalizing or law-finding science that somehow stands above the social sciences and history is not even appropriately debatable' (p. 195).

2   Geography is a lower-level science of spatial relations, applying in empirical contexts the laws of higher-level, generally more abstract sciences. (This may be a description of much of the geography of the 1960s.) This again seems to truncate the latter sciences and raises the question 'what differentiates economic geography from economics, and vice versa?'.

3   Geography is the study of geographical spatial relations. This implies that there are spatial phenomena not studied in the other social science disciplines and which can therefore be claimed as geography's: May can conceive of no objects which are purely geographical (or, in the parallel argument, purely historical either).

4   Geography is the study of 'things in reality' spatially. Yet again, this abstracts from other sciences, although May does admit that there are certain 'bits and pieces' which are not studied elsewhere; these, however, do not make a satisfactory empirical foundation for a separate discipline.

5   'Geography is not a generalizing or law-finding science of spatial relations' (p. 203).

The first four of these possibilities indicate that, because of the analytical indissolubility of time, space and matter, all social sciences are concerned with spatial relations. For May, as for Sack, therefore, geography cannot claim an independent status on the basis of the spatial variable. Moss (1970) reached a similar conclusion at about the same time:

> geometrical relationships must be assigned economic, social, physical, or biological meaning before they can in any sense become explanatory . . . though geometries may be important tools in geographical study and research, they cannot be a source of theory since their analogy with geographical phenomena is simply through particular logical structures, and not through explanatory deduction . . . such an application implies that space, area, distance, etc., are important in and of themselves, quite independently of any implications they may have in terms of diffusion, of cost, of time, or of process. This is manifestly false (p. 27)

Gregory (1978b, 1980) has criticized the extremely narrow, even superficial view of spatial processes which he perceives in the work of many spatial scientists (p. 106). Their view, he claims, is an instrumentalist one involving theories which cannot be validated conclusively but which can only be evaluated pragmatically against the real world. Bennett (1974), for example, accepts that his models do not mirror actual processes, but makes assumptions that they do allow policy formulation (and therefore produce self-fulfilling prophecies): because they can post-dict the world as it presently is, it is assumed that they explain it, and so can be used to predict the future.

The arguments reviewed in the present section are highly critical of much of the work undertaken by geographers in the fashion which dominated the

1960s, impelled by the 'Victorian myth of the supremacy of the natural sciences' (Gregory, 1978a, p. 21). The alternatives to such work are discussed in the next chapter.

Systems

The study of systems, as currently understood, was first introduced to the geographical literature by Chorley in 1962 although, as Foote and Greer-Wooten (1968) noted, systems analysis was promoted in Sauer's (1925) programmatic statement *The Morphology of Landscape* with the words 'objects which exist together in the landscape exist in interrelation': Garrison's (1960a) review paper made a similar point. To some extent, the adoption of a systems approach involved putting 'old' wine into new bottles', although the holistic approach currently espoused differs very much in format from that used by Sauer. More generally, the notion of a system has a long history, as Bennett and Chorley (1978, pp. 11–14) point out: teleological traditions, for example, postulate the world as 'a vast system of signs through which God teaches man how to behave' (p. 12), whereas functionalism links observed phenomena together as 'instances of repeatable and predictable regularities' of form.

The keystone of the study of systems is connectivity. As Harvey (1969a, p. 448) points out, reality is infinitely complex in its links between variables, but systems analysis provides a convenient abstraction of that complexity in a form which maintains the major connections. A system comprises three components (p. 451):

1 a set of elements;
2 a set of links (relationships) between those elements; and
3 a set of links between the system and its environment.

The last components may be non-existent, in which case the system is termed a closed one. Closed systems are extremely rare in reality, but are frequently created, either experimentally or, more usually in human geography, by imposing artificial boundaries, in order to isolate the salient features of a system. Thus, just as an internal-combustion engine comprises a set of linked elements which receives energy from its environment and returns spent fuel to that environment, so a set of settlements linked by communications networks forms a spatial system, with links to settlements outside the defined area of the system being the contacts with the environment. In a system, the elements have volumetric qualities and material flows along the links. As the system operates, so the various quantities may change.

The use of systems terminology to describe a spatial assemblage was widely adopted by human geographers in the 1960s, and it formed the basis, for example, of Haggett's (1965c) pioneering text. Not all applications of the terminology involved the study of explicitly spatial systems. In many cases, the elements were phenomena whose spatial locations were not studied, and the links between the phenomena were functional relationships. In such contexts, geographical study of causal systems was identical to the study of similarly structured problems in other disciplines, but that the phenomena occupied locations and that the links involved crossing space suggested the extra, geographical dimension which led to investigations of causal, spatial systems.

The early literature on systems analysis in geography was programmatic rather than applied; it suggested how the terminology might be applied in a research and teaching context, often reinterpreting old material (McDaniel and Hurst, 1968). Relatively few applications were reported, and more than a decade later, much of the literature assessed in a major review was written by other scientists (Bennett and Chorley, 1978). Nevertheless, Harvey (1969a) wrote that:

> If we abandon the concept of the system we abandon one of the most powerful devices yet invented for deriving satisfactory answers to questions that we pose regarding the complex world that surrounds us. The question is not, therefore, whether or not we should use systems analysis or systems concepts in geography, but rather one of examining how we can use such concepts and such modes of analysis to our maximum advantage (p. 479).

In the search for answers to this question, two variants on the systems theme have been employed. The first is *systems analysis*; the second is *general systems theory*, which is an attempt to provide a more unified science than current disciplinary boundaries allow. The two are sometimes confused, but will be treated separately here.

Systems analysis

Having defined a system of geographical interest, with all the attendant difficulties that this entails (Harvey, 1969a, pp. 445-9), how might it be studied? Several typologies of systems and systems analyses have been suggested.

Chorley and Kennedy (1971) identify four types of system (Figure 4.2). *Morphological systems* are statements of static relationships – of links between elements: they may be maps showing places joined by roads, or they may be equations describing the functional relationships between variables. Much of the spatial analysis described earlier in this chapter outlined such morphological systems. *Cascading systems* contain links along which energy passes from one element to another: factories can be portrayed as cascading systems, in that the output of one factory is in many cases the input for another. Each element may itself be a system (linked departments within a factory, for example), so a nesting hierarchy of cascading systems can be described, as with Haggett's (1965c) nodal regions and the input-output matrix representation of an economy (Isard, 1960): Berry (1966) has linked these two examples of cascading systems together in his inter-regional input-output study of the Indian economy. Within each element in a cascading system, the material flowing through is manipulated in some way (the industrial process in a factory, for example). The nature of the manipulative process may be ignored entirely in the investigation, with focus on the inputs and outputs only: in such case the representation of the element is termed a black box. White-box studies investigate the transformation process, whereas grey-box analyses make a partial attempt at their description.

Process-response systems are characterized by studies of the effects of linked elements on each other. Instead of focusing on form, as in the first two types, these are studies of processes, of causal interrelationships. In systems terms these may involve, for example, the effects of a variable X on another, Y; in the analysis of spatial systems they could involve the effect of variable X in place a on variable Y in place b, as with the effect of inflation in the United States on unemployment in the United Kingdom at some later date, perhaps, or with the transmission of a disease from one area of a country to another (Cliff *et al.*, 1975). Finally, there are *control systems*, which are special cases of process-response systems, having the additional characteristic of one or more key elements (valves) which regulate the system's operation and may be used to control it.

Attention has focused on the last two types. Langton (1972), for example, has suggested that process-response systems provide an excellent framework for the study of change in human geography. He identifies two sub-types. *Simple-action systems* are unidirectional in their nature: a stimulus in X produces a response in Y, which in turn may act as a stimulus to a further variable, Z. Such a causal chain is merely a reformulation of 'the characteristic cause-and-effect relation with which traditional science has dealt'

Figure 4.2 Types of system. In the diagrams, A, B and C indicate elements in the system, I = input, O = output, and in the control system, A is a valve. Source: Chorley and Kennedy (1971), p. 4).

(Harvey, 1969a, p. 455); in another language it is a process law.

More important, and relatively novel to human geography, is the second sub-type, *feedback systems*. According to Chorley and Kennedy (1971):

> Feedback is the property of a system or sub-system such that, when change is introduced via one of the system variables, its transmission through the structure leads the effect of the change back to the initial variable, to give a circularity of action (pp. 13–14).

Feedback may be either direct – A influences B which in turn influences A (Figure 4.3A) – or it may be indirect, with the impulse from A returning to it via a chain of other variables (Figure 4.3B). With *negative feedback* the system is maintained in a steady state by a process of self-regulation known as homeostatic or morphostatic: 'A classic example is provided by the process of competition in space which leads to a progressive reduction in excess profits until the spatial system is in equilibrium' (Harvey, 1969a, p. 460). But with *positive feedback* the system is characterized as morphogenetic, changing its characteristics as the effect of B on C leads to further changes in B, via D (Figure 4.3D).

The concept of feedback, with the associated notions of homeostasis and morphogenesis, gave 'the nuclei of the systems theory of change' (Langton, 1972, p. 145): as a consequence, Langton argues that the nature of feedback should be the focus of geographical study. In many spatial systems, feedback may be uncontrolled, but others may include a regulator, such as a planning policy (Bennett and Chorley, 1978). There are few geographical studies of such feedback processes, however. For homeostatic systems, Langton cites investigations of central place dynamics in which the pattern of service centres is adjusted as the population distribution changes, to reproduce the

Figure 4.3   Various types of feedback relationship within systems. Source: Chorley and Kennedy (1971, p. 14).

previous balance between supply and demand factors (e.g. Badcock, 1970). Morphogenetic systems are illustrated by Pred's (1965b) model of the process of urban growth, in which expansion in a sector generates, via a series of links, further expansion there, as in Myrdal's (1957) more general theory of cumulative causation: such systems modelling has been used to predict urban futures (Forrester, 1969). But in most of these studies the input from systems theory is slight, leading Langton (1972) to the paradoxical conclusions that:

> *First*, there is little correlation between the extent of the penetration of the terminology of systems theory and the rigorous application of its concepts. The 'empty' use of terminology, which is typified by the use of the term feedback as an explanatory device rather than as a description of a fundamental research problem, must be counter-productive in a situation in which the terms themselves may be given many subtle different shades of meaning. . . .
>
> *Second*, somewhat paradoxically, many of the concepts of systems theory are already used in geography without the attendant jargon and without apparently drawing direct inspiration from the literature of systems theory (pp. 159–60).

This second conclusion suggests that a discipline should develop the necessary conceptual framework itself without the introduction of relevant concepts from related disciplines, but Langton disputes this, arguing that systems theory clarifies questions in established theories, focuses directly on the processes of change, and forces careful analytical study.

One of the most substantial attempts to apply systems theory to a problem in human geography is the work of Bennett (1975) on the dynamics of location and growth in northwest England. Having represented the system – its elements, links, and feedback relationships – he estimated the influence of various external (i.e. national) events on the system's parameters, isolated the effects of government policy (Industrial Development Certificates) on the system's structure, and produced forecasts of the region's future spatio-temporal morphology. He has developed the forecasting aspects of this methodology in later papers (Bennett, 1978a, 1979), and has suggested how an optimal distribution of government grants can be achieved (Bennett, 1981b).

One type of study which has firmly adopted the systems approach covers the border area between human and physical geography. The *ecosystem* is a process-response system concerned with the flows of energy through biological environments, most of which include, or are affected by, man. It is also a control system in that the living components act as regulators of the energy flows: 'they further represent a major point at which human control systems must intersect with the natural world' (Chorley and Kennedy, 1971, p.330). Most naturally occurring ecosystems are, for much of the time, homeostatic (see Chapman, 1977, Chapter 7), but human entry often transforms them into morphogenetic systems, with potentially catastrophic effects.

Stoddart (1965, 1967b) has argued that the ecosystem should be employed as a basic geographical principle, but despite other programmatic statements (e.g. Clarkson, 1970), including two based on the associated concept of community (Morgan and Moss, 1965; Moss and Morgan, 1967), Langton's conclusion about relatively little substantive research would appear to be valid (see Grossman, 1977). Similar attempts, but involving less consideration of the biotic environment, have been made in allied disciplines, and have occasionally been imported into the geographical literature. The human ecosystem models of sociologists (e.g. Duncan, 1959; Duncan and Schnore, 1959) have been used as frameworks for the investigation of migration (Urlich, 1972) and of urbanization (Urlich Cloher, 1975), for example, and the operational research techniques of economists, with their important feedback mechanisms, have stimulated some work in transport geography (e.g. Sinclair and Kissling, 1971).

The most comprehensive attempt to forge a systems approach to geographical study has been provided by Bennett and Chorley (1978). The intention of their book is to provide 'a unified multi-disciplinary approach to the interfacing of "man" with "nature" ' (p. 21), with three major aims:

> First, it is desired to explore the capacity of the systems approach to provide an inter-disciplinary focus on environmental structures and techniques. Secondly, we wish to examine the manner in which a systems approach aids in developing the interfacing of social and economic theory, on the one hand, with physical and biological theory, on the other. A third aim is to explore the implications of this interfacing in relation to the response of man to his current environmental dilemmas. . . . It is hoped to show that the systems approach provides a powerful vehicle for the statement of environmental situations of ever-growing temporal and spatial magnitude, and for reducing the areas of uncertainty in our increasingly complex decision-making arenas (p. 21).

This mammoth task (undertaken in a mammoth book of 624 pages!) involves elucidation of not only the 'hard systems' of physical and biological sciences but also the 'soft systems' characteristic of the social sciences. With regard to the latter, they cover a very large and fertile literature concerned, first, with the cognitive systems describing man as a thinking being and the decision-making systems used by humans, as individuals and in groups, and, secondly, with the socio-economic systems made up of very many of these interacting individuals and groups. They then attempt to interface the two types, because:

> in large-scale man-environment systems the symbiosis of man as part of the environment of the system he wishes to control introduces all the indeterminacies of socio-economic control objectives. . ... In particular, we need to ask what are the political and social implications of control, and for whom and by whom is control intended? (p. 539)

Not surprisingly, they end their essay into the field with a discussion of the very many substantial problems involved in such interfacing, although many of these are firmly tackled in the text (which was written after the widespread introduction to the literature of many of the criticisms discussed in the next two chapters).

This use of systems analysis is based on the assumption (usually implicit) underlying much positivist work in human geography, that valid analogies can be drawn between human societies on the one hand and both natural phenomena-complexes and machines on the other. Individual elements in a system have predetermined roles, and can act and change in certain ways only – depending on the structure of the system and its interrelationships with the environment. As a descriptive device, this analogy allows the structure and operation of society and its components to be portrayed and analysed, and it provides a source of ideas from which hypotheses can be generated (see Coffey, 1981). And once a system has been defined and modelled, systems analysis can be used as a predictive tool, to indicate the nature of the elements and links following certain environmental changes (such as the introduction of new elements and/or links, as in the classic Lowry model used to predict the impact of new land-use configurations on traffic flows: Batty, 1978).

The potential fertility of this analogy has been examined by several authors. Wilson (1981b), for example, has discussed methods of analysing environmental systems in which he defines environment as 'natural, man-nature, and manufactured 'systems of interest' ' (p. xi), in all of which

> the main concern is with *complicated* systems whose components exhibit high degrees of interdependence. The behaviour of the 'whole' system is then usually something very much more than the sum of the parts (p. 3).

His case is that moorland ecosystems, water resource systems, and cities (his three initial examples) can all be studied in the same way and his book investigates the methods for this; a further book (Wilson, 1981a) presents the mathematics for studying systems in which the rates and directions of change suddenly alter. Huggett (1980) similarly argues that systems analysis has wide applications in both human and physical geography, plus the interface where man and environment interact, and his book illustrates the methods of systems analysis (see also Huggett and Thomas, 1980). And once the system has been successfully modelled, it can then be manipulated using control theory which:

> is a *dynamic* optimization technique . . . [which] permits optimal allocation over long time horizons . . . [and] shifts emphasis from mere model construction to model use (Chorley and Bennett, 1981, p. 219).

Such a combination of models describing systems with a theory of system control has a wide range of potential applications, according to Chorley and Bennett, in such fields as pollution control, catchment management, inter-area resource allocation and urban planning. It suggests a commonality of interest, focused on methods, between applied physical and applied human geography.

Systems theory, information, and entropy

The definition of a system presented so far in this chapter is of a series of linked elements interacting to form an operational whole. This has been challenged by Chapman (1977) who opens his book with the statement:

> I do not think that the concept of a system will have any great operational consequences in geography for a long time yet. It represents an ideal that the real world does not fully approach. On the other hand, in conceptual terms I think the concept is extremely important and useful, and that it has a great and immediate role to play for those who are about to plan the strategy of their research. As a framework for analysis, it has no current peers (p. 6).

For Chapman, a system comprises a series of elements which can take alternative states, and his definition − following Rothstein (1958) − is

> A system is a set of objects where each object is associated with a set of feasible alternative states: and where the actual state of any object selected from this set is dependent in part or completely upon its membership of the system. An object that has no alternative states is not a functioning part but a static cog (p. 80).

An example of such a system is a number of farms, each comprising a series of fields: every farmer has to decide how to use each of his fields. In part, each of his decisions will reflect his general operations and the uses to which he puts all of his other fields; in part, it will be a function of the external market and the decisions made by other farmers regarding their own fields. Thus there is a large number of possible states for the system of fields − different configurations of land uses. According to Chapman, systems analysis should involve investigation of these configurations and the placing of the observed pattern within the context of the alternatives:

> to theorize merely about what does exist is not very useful. If we restrict ourselves to that alone, all explanation will be merely historical accidental. At all stages it is most important to include consideration of what else could have been. The definition of organization in a system even explicitly requires the assessment of what else could have been (pp. 120–1)

Taylor and Gudgin (1976) have argued in the same way in a particular research context: instead of simply asking 'is there a bias in the electoral districting of a borough?' they ask 'what is the likelihood of a bias occurring, given the constraints of the system?', and they have set the study of electoral districting and its consequences on a strong statistical footing (Gudgin and Taylor, 1979).

Such analyses focus on one configuration of the system as a sample from a set of possible configurations. One of the key concepts is that of *entropy*. In general terms, the entropy of a system is an index of uncertainty. (In the second law of thermodynamics an increase in system entropy is an increase in system uncertainty. A good example of this is the introduction of a layer of hot water onto a body of cold water. Initially, the two types are separate, and one can be completely certain about the location of the hot molecules, for example. But with no external influence the two slowly mix, until all are at the same temperature. As the mixing proceeds, so the entropy increases.)

Social scientists have drawn their usage of entropy from two separate, though linked, definitions. Thermodynamic entropy relates to the most probable configuration of the elements, within the constraints of the system's operations. In information theory, entropy refers to the distribution of the elements across a set of possible states, and is an index of element dispersion. One can be completely certain about a distribution, in terms of predicting where one element will be, if all elements are in the same state; conversely, one will be most uncertain when elements are equally distributed through all possible states, so that prediction of the location on any one element is most difficult. (An example of this use is in Johnston, 1976a; see also Webber, 1977, on the relationship between entropy and uncertainty, and Thomas, 1982.)

Used in its simplest form, the information-theory measure of entropy is another descriptive index, but it can be developed in a variety of ways. Chapman (1977) illustrates three uses:

1 as a series of indices of variations in population distributions;
2 as an index of redundancy in a landscape, where redundancy is defined as relating to a regular sequence so that it is possible to predict the land use at place a, for example, from knowledge of the land uses at neighbouring places; and
3 as a series of measures of reactions to situations in states of uncertainty.

In general terms, although not in detailed methodology, these continue the tradition of Stewart and Warntz's macrogeography (p. 67): the aim is to describe a pattern rather than to explain it, although the nature of the constraints used to derive the entropy measures provides an input to explanatory modes of analysis (Webber, 1977).

The usage of entropy as developed in statistical mechanics rather than in information theory was introduced to the geographical literature by Wilson (1970): His initial example was a flow matrix. The number of trips originating in a series of residential areas is known, as is the number ending in each of a series of workplace areas, but the entries in the cells of the matrix – which people move from which residential area to which workplace area – are unknown. What is the most likely flow pattern? Even with only a few areas and relatively small numbers of commuters, the number of alternatives is very large. Wilson defines three states of the system. The first is the macro-state, comprising the number of commuters at each origin and the numbers of jobs at each destination. The second is the meso-state, comprising a particular flow pattern: five people may go from zone A to zone X, for example, and three from the same origin to zone Y, but it is not known which five are in the first category and which three in the second. A micro-state, on the other hand, is a particular example of a flow pattern – one of the many possible configurations of eight people moving from zone A, five to X and three to Y. Entropy-maximizing procedures find that meso-state with the largest number of micro-states associated with it; in other words it is the meso-state about which the analyst is least certain of which configuration produces the pattern. Such a procedure shows that

> the most probable distribution is that with the greatest number of micro-states giving rise to it. Thus the distribution corresponds to the position where we are most uncertain about the micro-state of the system, as there are the largest possible number of such states and we have no grounds for choosing between them (p. 6).

This too follows the macrogeography tradition. It is not an attempt at explanation. Rather Wilson sees his work as illustrating

> the application of the concept of entropy in urban and regional modelling, that is, in *hypothesis development*, or theory building. ('Model' and 'hypothesis' are used synonymously, and a theory is a well tested hypothesis.) . . . the entropy-maximizing procedure enables us to handle extremely complex situations in a consistent way (pp. 10–11).

The hypothesis that the entries in the flow matrix conform to the most likely distribution can be tested against 'real' data. If it is falsified, either entirely or in part, it can be refined by building in more constraints. Wilson does this with his intra-urban transport models, for example, by introducing travel-cost constraints, different types of commuters (class, age, etc.), different types of jobs, and so on. The aim is to describe the most likely system structure from a given amount of information which is incomplete.

Wilson has developed both the theory of his modelling and the substantive applications (see Wilson, 1981b). His general text (Wilson, 1974) contains a whole family of models that can be used to represent, and then to forecast, the various components of a complex spatial system such as an urban region. Some of these models have been expanded (e.g. Rees and Wilson, 1977, on demographic accounts) and they have all been applied, with varying success, to the West Yorkshire region (Wilson, Rees and Leigh, 1977).

Wilson's work has been applied and developed by others (see Batty, 1976, 1978, for example). In reviewing his 1970 presentation, Gould (1972) termed it 'the most difficult I have ever read in geography' but continued 'he has planted a number of those rare and deep concepts whose understanding provides a fresh and sharply different view of the world' (p. 689). Webber (1977) has extended Wilson's argument that the purpose of entropy-maximizing models is to draw conclusions from a data set which are 'natural' in that they are functions of that data set alone and contain no interpreter bias. For him, they provide a convenient way of oganizing processes of thinking about a complex world, and he identifies an 'entropy-maximizing paradigm' (p. 262) which focuses on location models (the probability of an individual being in a particular place at a particular time), on interaction models (the probability of a particular trip occurring at a particular time), and on joint location/interaction models.

In emphasizing aggregate patterns, this work is macrogeographic:

according to Webber (1977) 'The entropy-maximizing paradigm asserts . . . that, though the study of individual behavior may be of interest, it is not necessary for the study of aggregate social relations' (p. 265). The patterns predicted by the models are functions of the constraints (which are the information provided at the meso-state), so that knowledge of these means that 'the entropy-maximizing paradigm is capable of yielding meaningful answers to short-run operational problems' (p. 266) and thus is of immense value for immediate planning purposes. But:

> in the longer run, much of the economic system is variable: the constraints and the spatial form of the urban region may change . . . the research task facing entropists is (1) to identify the constraints which operate upon urban systems, which is partly an economic problem; (2) to deduce some facets of the economic relations among the individuals within the system from the use of the formalism; and (3) to construct a theory which explains the origins of the constraints. Only when the third task has been attempted may the paradigm be adequately judged (p. 266).

Thus the entropy-maximizing model acts not only as a 'black box' forecasting device (p. 113) but also as a hypothesis: if the operation of the systems described is to be understood, the axioms – the constraints – must themselves be explained. Given the nature of the constraints (in Wilson's initial example, why people live where they do, why people work where they do, and why they spend a certain amount of time, money and energy on transport) the task is a major one: entropy-maximizing models aim to clarify it and indicate the most fruitful avenues for investigation.

The theory of systems and general systems theory

Like regional science, the development of General Systems Theory (GST) has been very much tied up with the academic career of one man – in this case Ludwig von Bertalanffy (see von Bertalanffy, 1950). It reflects an attempt to unify science via perspectivism instead of the more usual division of science through reductionism. Its focus is on isomorphisms, the common features among the systems studied in different disciplines, and 'Its subject matter is the formulation and derivation of those principles which are common for systems in general' (Walmsley, 1972, p. 23). The goal is a metatheory with rules that apply in a variety of contexts; application (which might be termed the theory of systems rather than GST) is usually by analogy from one discipline in order to advance understanding in another (as in Chappell and Webber's, 1970, use of an electrical analogue of spatial diffusion processes). For geography, GST offers an organizing framework; GST itself is an empirical exercise using inductive procedures to fashion general theories out of the findings of particular disciplines (see Coffey, 1981).

It has been claimed that there have been no advances in either the theoretical base or the empirical application of GST (Greer-Wootten, 1972). It has been employed by some geographers, however: Woldenberg and

Berry (1967) drew analogies between the hierarchical organization of rivers and of central-place systems, for example; Berry (1964a) argued that cities are open systems in a steady-state, as exemplified by the stability of their behaviour-describing equations; and several authors (e.g. Ray, Villeneuve and Roberge, 1974) have applied the concept of allometry – that the growth rate of a component of an organism is proportional to the growth of the whole – in several contexts.

According to its proponents, the advantages of GST to human geography lie in its interdisciplinary approach, its high level of generalization, and its concept of the steady-state of an open system (Greer-Wootten, 1972; Walmsley, 1972), but they also contend that geography's strong empirical tradition means that it has more to contribute than to take from GST. But one critic has claimed that 'General systems theory seems to be an irrelevant distraction' (Chisholm, 1967, p. 51), an argument based largely on a paper by Chorley (1962) referring to the Davisian system of landscape development. Chisholm summarizes the case for GST as:

1   there is a need to study systems rather than isolated phenomena;
2   there is a need to identify the basic principles governing systems;
3   there is value in arguing from analogies with other subject matter; and
4   there is a need for general principles to cover various systems.

In his view, however, something as grand as a metatheory is unnecessary in order to convince people of the need to understand what they study, of the value of interdisciplinary contact, and of the potential fertility of arguing by analogy.

Conclusions

As stressed in the previous chapter, much of the force of the developments in human geography during the 1950s and 1960s related to methodology and the approach to traditional geographical questions. In addition, however, attempts were made to inaugurate and press a particular geographical point of view; the two main themes advanced have been reviewed here. The first – the spatial-science viewpoint – was fairly widely and rapidly accepted and many geographers placed the spatial variable at the centre of their research efforts: as will be indicated in the next chapter, their work came under increasing attack after the mid 1960s. The second – the systems approach – has received much less detailed attention, despite frequent gestures of approval towards it. Compared with the spatial-science view, which could be rapidly assimilated within the developing statistical methodology (although note the critique discussed on p. 107), the systems approach was technically much more demanding, and perhaps for that reason attracted fewer active researchers. These have remained the source of fertile ideas, however, and have continued to publish major works throughout the 1970s, paralleling, and in some cases (e.g. Bennett and

Chorley, 1978) responding to, the academic movements outlined in the following chapters.

Human geography as spatial science was inaugurated in North America in the 1950s. By the end of the 1960s it was dominating work there and in the rest of the English-speaking world. Most research was positivist in its tone, if not its detail, seeking to describe patterns of spatial organization and to account for these as consequences of the influence of distance on human behaviour. A great deal of that work was quantitative, and it contributed to bodies of theory, either about spatial organization in general or certain aspects of it in particular (as with industrial location: Smith, 1981). And, as illustrated here, it has influenced other academic disciplines and the planning profession.

Since the 1970s, positivist spatial science has been under considerable attack among Anglo-American human geographers, as discussed in the next two chapters. But this attack has had little effect on the volume of work done in this 'paradigm', as illustrated by a major review of *Quantitative Geography* by British workers (Wrigley and Bennett, 1981); this work has been stimulated and aided by advances in data handling technology and display (Rhind, 1981). At the same time, the approach has been fostered outside the Anglo-American realm, especially in Western Europe where the reaction to quantification has been stronger than in Britain and North America (see Bennett, 1981c).

This great volume of work suggests that there has been 'progress' within the positivist paradigm. Whether this is so is not easy to assess. Certainly, as suggested here, there have been substantial advances in the development and adaptation of sophisticated modelling and analytical techniques (on the former, see Papageorgiou, 1976), although to some these advances, although 'scholarship of the highest order', may reflect 'a misdirection of effort' (King, 1979b, p. 157) because they have not illuminated spatial organization as much as they might. But new techniques continue to be advanced (Beaumont and Gatrell, 1982). Empirical work in the spatial science context has very much taken the form of case studies set in a general theoretical context, and – despite the efforts of some general surveys (e.g. Haggett, Cliff and Frey, 1977) – it is not clear whether such *case studies* have provided any more *general* understanding than has been so in historical and cultural geography, which continued to operate quietly through individual investigations throughout the period under discussion here. What is without doubt, however, is the volumetric impact of spatial science on human geography.

One of the most salient features of this conception of human geography is clearly its focus on space, especially distance as a component of space, as the central element of geographical study. As such, it presents a particular view of the spatial variable. Sack (1980) has indicated that this is a very partial presentation. He identifies two main treatments of space in society.

1 The sophisticated-fragmented treatment, which handles space as part of

a defined methodology. This can be subdivided into three. First, there is the physical science treatment of space, which focuses on objective elements. Secondly, there is the largely separate, humanities treatment which – through prose, verse, works of art, etc. – handles space subjectively. Thirdly, there is the social science treatment, which overlaps the other two. All three are sophisticated, because the analyses by physical scientists, artists, and social scientists do not influence (at least directly) that which they are depicting; they are fragmented because they deal with parts only.

2 The unsophisticated-fused treatments, characterized by the practical, the child's, and the mythical-magical views of space. These are unsophisticated because they are not abstract, and fused because they encompass facts and feelings and do not seek to separate out elements. Perhaps most importantly, they are different because they involve intense feelings that are relevant to activity in the world.

Spatial science, as defined here, covers only the overlap area between the physical and social sciences in the sophisticated-fragmented treatment. Interest in the overlap area with the humanities, and in the unsophisticated-fused treatment, has led to the consideration of alternative formulations of human geography, as outlined in the next chapter.

5

Behavioural geography and alternatives to positivism

The changes outlined in the previous two chapters had been widely adopted by the mid 1960s, particularly in the United States, in Canada, and in Great Britain, but also in those other countries whose academic traditions and life are closely tied to the North Atlantic realm. As already documented, the changes were, not surprisingly, resisted by defenders of previous methodologies and philosophies. Almost contemporaneous with such defences were attacks on the positivist approaches, attacks based not on the growing split between the 'old' and the 'new' in human geography but rather on the perceived failures of the latter to provide viable modes of understanding. Some of the attacks came from without, from those who had neither accepted nor operated the tenets of positivist spatial science, but in part also they came from within, from those who had tried the 'new' and were disappointed with their 'accomplishments'. There was some contact between the two groups, including cross-referencing of literature, but their arguments led in different directions. The initial attack from within led to a modification of approach to theory, but not method, whereas the attack from without suggested new methodologies and philosophies. Separate reviews of the two are presented here.

Towards a more positive, behaviouristic spatial science

The main ground for disillusion within the positivist camp was a growing realization that the models being propounded and tested were not very good descriptions of reality, so that progress towards the development of geographical theory was painfully slow. Thus the large body of work based on central place theory, for example, was built on certain axioms regarding human behaviour, with regard to choice between spatial alternatives, and from these axioms a settlement pattern was deduced. But the deductions were often only vaguely reflected in settlement morphologies, which suggested that the axioms on which they were based provided a weak foundation for understanding this aspect of the spatial organization of society. The theory suggested how the world would look under certain circumstances of economic rationality in decision-making; that those circumstances did not prevail suggested that the world should be looked at in other ways in order to understand how people do behave and structure their spatial organization. As Brookfield (1964) put it, with regard to the whole family of models then popular:

> We may thus feel that we have proceeded far enough in answering our questions when, by examination of a sufficient number of cases, we can make assertions such as the following: population density diminishes regularly away from metropolitan centres in all directions; crop yields diminish beyond a certain walking distance from the centres of habitation; air-traffic centres lying in the shadow of major centres do not command the traffic that their populations would lead us to expect. . . . Such answers, which represent the mean result of large numbers of observations whether statistically controlled or otherwise, are valuable in themselves, and sufficient for many purposes. But each is also an observation demanding explanations which may seem self-evident, or which may in fact be very elusive. Furthermore, there will be exceptions to each generalization, and in many cases there are also limits to the range of territory over which they hold true. Both the exceptions and the limits demand explanation (p. 285).

The issues being raised, therefore, were concerned not with the basic goal of positivist work – the establishment of generalizations and theories – but rather with the particular route being taken to that goal. The criticisms focused on the models being used, and their axioms regarding rational economic behaviour. The hypotheses derived from these models were, at best, being only weakly verified in empirical studies. The need was for better models, and the search for these took a more inductive route than that previously followed. Rather than base hypotheses on assumptions about behaviour, behaviour was to be investigated inductively, to provide inputs to a superior range of models. (This involved creating a new 'image of real world structure': Figure 3.1.)

Rationality in land-use decisions

One of the first attempts by geographers to explore behaviour inductively, to provide inputs for later modelling, was the series of investigations into human responses to environmental hazards, initially floods, which was organized at the University of Chicago during the late 1950s and early 1960s. Its director was Gilbert White, whose own thesis on human adjustment to floods was published in 1945. His associates developed a behaviourist approach for studying reactions to the hazards, basing this on Herbert Simon's (1957) theories of decision-making. Thus Roder (1961), for example, categorized Topeka residents according to their attitudes to the probability of future floods there, concluding that:

> Flood danger is only one of the variables affecting the choices of the flood-plain dweller, and many considerations operate to discourage a resident from leaving the flood plain, even when he is aware of the exact hazard of remaining (p. 83).

Such behaviour, it seemed, did not fit easily into the notions of profit-maximizing decision-making on which geographical theories were currently being built.

A major exponent of the behaviouristic approach was Kates (1962), who began his study of flood-plain management with the statement that 'The way men view the risks and opportunities of their uncertain environments plays a significant role in their decisions as to resource management' (p. 1). In studying such decision-making, Kates developed a schema which he claimed was relevant to a wide range of behaviours. It was based on four assumptions.

1 Men are rational when making decisions. Such an assumption may be either prescriptive – describing how men should behave – or descriptive of actual behaviour. The latter seems to be the most fruitful, both for understanding past decisions and for predicting those yet to be made. Kates suggested adoption of Simon's concept of bounded rationality as a basis for such study. According to this, decisions are made on a rational basis, but in relation to the environment as it is perceived by the decision-maker, which may be quite different from either 'objective reality' or the world as seen by the researcher.

> In this model, men bounded by inherent computational disabilities, products of their time and place, seek to wrest from their environments those elements that might make a more satisfactory life for themselves and their fellows (p. 16).

Rational decision-making is constrained, therefore, and is not necessarily the same as the maximum rationality assumed in the neo-classical normative models discussed in earlier chapters of this book.

2 Men make choices. Many decisions are either trivial or are habitual so that they are accorded little or no thought immediately before they are made. Some major decisions regarding the environment and its use may also be habitual, but such behaviour usually only develops after a series of conscious choices has been made, which leads to a stereotyped response to similar situations in the future.

3 Choices are made on the basis of knowledge. Only very rarely can a decision-maker bring together all of the information relevant to his task, and frequently he is unable to assimilate and use all that he has.

> Thus, a descriptive theory of choice must deal with the well informed and the poorly informed and the choices that men make under certainty, risk or uncertainty . . . such a theory must deal with the eventuality that not only do the conditions of knowledge vary, but the personal perception of the same information differs (p. 19).

4 Information is evaluated according to predetermined criteria. In habitual choice the criterion is what was done before, but in conscious choice the information must be weighed according to certain rules. Some normative theory prescribes maximizing criteria (of profits, for example); descriptive theory may use Simon's notion of satisficing behaviour, involving decision-makers who seek a satisfactory outcome (a given level of profit, perhaps) only.

The results of decision-making which do not match the predictions from the sorts of theories being used by spatial scientists do not necessarily imply irrational behaviour, therefore. Instead, most decisions are made rationally on the basis of a, probably non-random, selection of information, are intended to satisfy some goal which is not to make the most perfect decision, and are based on criteria which vary somewhat from individual to individual. Having learned a satisfactory solution to a given class of problems, decision-makers will then continue to apply it every time such a problem occurs, unless changed circumstances require a re-evaluation.

Kates's study aimed to understand why people choose to live in areas which are prone to flooding. Their information was based on their knowledge and experience, and they could be scaled according to the certainty of their perceptions regarding further floods. In justifying their decisions, most were boundedly rational, and had made conscious choices in order to satisfy certain objectives. Similar work was reported by a number of others associated with White's leadership, and, as he points out, their findings had some impact on public policy formulation in the United States (White, 1973). The work was extended to cover a wide range of environmental hazards, providing inputs for both national policies and programmes of inter-national cooperation (Burton, Kates and White, 1978). Their initial impact on the wider geographical enterprise was not great, however, especially in the early years of their work. This was perhaps because they were operating on the boundaries between human and physical geography, which few American researchers approached, and it was probably the later work of others which brought Simon's ideas more forcefully before the geographical audience.

Wolpert and the decision process in a spatial context

For many human geographers, it was probably a paper published by Julian Wolpert in the *Annals of the Association of American Geographers* for 1964 which introduced them to the behaviourist alternative to the normative approaches then popular. (Wolpert's paper was based on his PhD thesis submitted to the University of Wisconsin, and is further testament to the innovative qualities of the geographers there, and of their contacts with other social scientists, including agricultural economists. His research was conducted in Sweden, another hearthland of geographical innovations.) The normative theory espoused by many geographers at that time assumes an economic man who, according to Wolpert (1964),

> is free from the multiplicity of goals and imperfect knowledge which introduce complexity into our own decision behaviour. Economic Man has a single profit goal, omniscient powers of perception, reasoning, and computation, and is blessed with perfect predictive abilities . . . the outcome of his actions can be known with perfect surety (p. 537).

In the study of spatial patterns, however,

> Allowance must be made for man's finite abilities to perceive and store informa-
> tion, to compute optimal solutions, and to predict the outcome and future events,
> even if profit were his only goal (p. 537).

Thus farmers face an uncertain environment – both physical and economic – when they make their individual land-use decisions, which in
aggregate comprise the land-use map. Wolpert suggested that differences
between these decisions and those that would be made by economic man
should reflect aspects of the farmers' economic and social environments.

Comparing the labour productivity of farms in an area of Sweden with
what could have been achieved under optimizing decision-making, Wolpert
decided that the farmers were probably satisficers, although such a
hypothesis is difficult to verify without detailed knowledge of aspiration
levels. How they acted was undoubtedly contingent upon their available
information, and clear spatial variations in the levels of potential productivity achieved suggested parallel spatial variations of knowledge. Only con-
spicuous alternatives are considered, it was suggested, and the result is
rational behaviour, adapted to an uncertain environment.

Wolpert (1965) continued this theme with studies of migration, aiming to
model the decision-making which lies behind the patterns of migration
reported in census volumes and assiduously analysed by spatial scientists.
To him, the gravity model is inadequate as a representation of such flow
patterns; indeed 'Plots of migration distances defy the persistence of the
most tenacious of curve fitters' (p. 159). Boundedly rational man makes
decisions, first, whether to move, and secondly, where to, and he does so on
the basis of his place utilities, his evaluations of the degree to which each
location, including that which he currently occupies, meets his defined
needs. Not only is the information on which he bases these utilities far from
complete; for many places he has none. Thus each individual has an action
space – 'the set of place utilities which the individual perceives and to
which he responds' (p. 163) – whose contents may deviate considerably
from that portion of the 'real world' which it purports to represent. Once the
first decision – to migrate – has been made, then the action space may be
changed as the potential mover searches through it for potential satisfactory
destinations and, if necessary, extends the space if no suitable solution to the
search can be found.

Wolpert's papers heralded – certainly in timing and to some extent in
influence too – the development of what was termed behavioural geography
(Cox and Golledge, 1969) 'united by a concern for the building of geographic theory on the basis of postulates regarding human behaviour . . .
upon social and psychological mechanisms which have explicit spatial correlates and/or spatial structural implications' (pp. 1–3). This behavioural
approach has focused on topics related to decision-making in spatial contexts. Golledge (1969, 1970), for example, has looked at models of learning
about space and of habitual behaviour, and with Brown has investigated
methods of spatial search (Golledge and Brown, 1967). Others have

researched into information flows, on which decisions are based, indicating the influence of local context on behaviour (Cox, 1969), and Brown and Moore (1970) have extended Wolpert's place utility and action space concepts for the study of intra-urban migration.

The aim in behavioural geography, according to a review by Golledge, Brown and Williamson (1972), has been to derive alternative theories to those based on economic man, 'more concerned with understanding why certain activities take place rather than what patterns they produce in space' (p. 59), which involves 'the researcher using the real world from a perspective of those individuals whose decisions affect locational or distributional patterns, and . . . trying to derive sets of empirically and theoretically sound statements about individual, small group, or mass behaviours' (p. 59): man is an active decision-maker, not a reactor to particular institutionally—created stimuli; (Cox and Golledge, 1981; Thrift, 1981). In their evaluation of such behaviouristic endeavours, Golledge, Brown and Williamson indicate the seminal influence of Hägerstrand (1968), who used the concept of the mean information field (analogous to the action space of a place's residents) to model migration flows and the adoption of innovations. The initial interest in resource management decisions, as outlined above, was followed by an extension from environmental perception and decision-making into aspects of attitudes and motivation. These were applied to studies of migration, the diffusion of innovations, political behaviour — especially voting, perception, choice behaviour, and spatial search and learning. By studying behavioural processes in these contexts, the aspiration was to increase geographers' understanding of how spatial patterns evolve, thereby complementing their existing ability to describe such patterns. Morphological laws and systems are insufficient of themselves for understanding; the amalgamation of concepts about decision-making taken from other social sciences with geography's spatial variable would allow development of process theories that could account for the morphologies observed.

Further developments include another series of pioneering papers by Wolpert and his associates, this time relating to political decision-making. Regarding the distribution of certain artefacts in the landscape, Wolpert (1970) pointed out that the location of, for example, a public facility in an urban area frequently is the product of policy compromise:

> Sometimes the location finally chosen for a new development, or the site chosen for a relocation of an existing facility, comes out to be the site around which the least protest can be generated by those displaced. Rather than being an optimal, a rational, or even a satisfactory locational decision produced by the resolution of conflicting judgements, the decision is perhaps merely the expression of rejection by elements powerful enough to enforce their decision that another location must not be used . . . These artefacts are rarely 'the most efficient solutions', and frequently not even satisfactory neither for those responsible for their creation nor for their users (p. 220).

(This argument avoids considering any definition of optimal, in either or both of economic and political terms.) Such decisions involve what Wolpert terms maladaptive behaviour. The work of Kates and others had suggested that decisions are adaptively rational, within the constraints of uncertainty, utility, and problem-solving ability. Coping strategies under the mutual exchange of threats between interested parties, however, can lead to decision-making which does not involve the careful and methodical investigation of alternatives until a satisfactory solution is found. Instead decisions are the consequences of conflict between groups with different attitudes and motivations, and are not the result of joint application of criteria on whose relevance there is a consensus.

> This formulation then lays the framework for the interpretation of locational decisions which appear to be more the product of stress responses than the end result of a dispassionate and considered selection of alternatives posited by the classical normative approaches or even the Simon scheme of bounded rationality (p. 224).

Wolpert and his associates have applied this formulation in a variety of contexts, such as the routes for intra-urban freeways and the siting of community mental-health facilities (Wolpert, Dear and Crawford, 1975).

Mental maps

One aspect of behavioural analysis enthusiastically adopted by a number of workers was the concept of mental map of the environment which guides the deliberations of decision-makers. The term mental map was not new to the geographical literature, having been used by Wooldridge (1956) in his descriptions of the perceived environments within which farmers make their land-use decisions. The seminal paper which revived it was published by Gould in 1966; his guiding belief was that

> If we grant that spatial behavior is our concern, then the mental images that men hold of the space around them may provide a key to some of the structures, patterns and processes of Man's work on the face of the earth (p. 182 – reference to 1973 reprint).

Increasingly, he argued, location decisions are being taken with regard to perceived environmental quality, so it is necessary to know how people evaluate their environment, and whether their views are shared by their contemporaries. To investigate such questions, Gould required respondents in various countries to rank-order places according to their preferences for them as places in which to live, and these rankings were analysed to identify their common elements – the group mental maps (Gould and White, 1974). Such maps, it was argued, are useful not only in the analysis of spatial behaviour but also in the planning of social investment – such as differential salaries to attract people to less desirable areas.

Those who followed Gould's lead investigated a range of methods for identifying and analysing spatial preferences (Pocock and Hudson, 1978).

The results obtained provided little input to theoretical development, however, and Downs (1970) wrote that

> Even the most fervent proponent of the current view (that human spatial behaviour patterns can partially be explained by a study of perception) would admit that the resultant investigations have not *yet* made a significant contribution to the development of geographic theory (p. 67).

Apart from Gould's rank-ordering procedures, Downs identified two other major approaches to the study of environmental images: the structural approach, which inquires into the nature of the spatial information stored in people's minds and which they use in their everyday lives – Lynch's (1960) book was a model for such work – and the evaluative approach, in which (Downs, 1970) 'The question is, what factors do people consider important about their environment, and how, having estimated the relative importance of these factors, do they employ them in their decision-making activities' (p. 80). With this evaluative approach geographers moved into the wider field of cognitive mapping – 'a construct which encompasses those cognitive processes which enable people to acquire, code, store, recall and manipulate information about the nature of their spatial environment' (Downs and Stea, 1973, p. xiv). In this they worked alongside both psychologists, who were becoming increasingly interested in man's relationship to a wider area than his proximate environment and in the development of relevant non-experimental research techniques, and designers concerned with the creation of more 'liveable' environments. The journal *Environment and Behavior* was launched in 1969 to cater for this interdisciplinary market, but despite some interest (e.g. Tuan, 1975a; Downs and Stea, 1977; Pocock and Hudson, 1978; Porteous, 1977) the general field has not made a major impact within human geography.

The concept of 'mental map' and the associated process of 'cognitive mapping' – that 'seems to imply the evocation of visual images which possess the kinds of structural properties that we are familiar with in "real" cartographic maps' (Boyle and Robinson, 1979, p. 60) – have become the centre of considerable debate both among behavioural geographers and between them and outside critics. Gould's initial work, and that which it stimulated, has been criticized as the study of space preferences only (Golledge, 1981a; see also Golledge, 1980, 1981b; Guelke, 1981; Robinson, 1982). But, as Downs and Meyer (1978) make clear, 'perceptual geography' – 'the belief that human behaviour is, in large part, a function of the perceived world' (p. 60) – extends much further than the elicitation and mapping of space preferences. The fundamental argument of behavioural geography is that: (i) environmental images exist; (ii) these images can be identified accurately (as in the evaluation of landscapes: Penning-Rowsell, 1981); and (iii) 'there is a strong relationship between environmental images and actual behaviour' (Saarinen, 1979, p. 465). The nature of these images – whether they are maps in the understood sense of that word and whether they can be apprehended by researchers – remains a problem: the concept of a 'mental map' may be a red herring but, to behavioural geographers, the argument on which it is based is not.

Time-space geography

During the late 1960s, Hägerstrand began to formulate a framework for the study of human geography based on a joint time-space set of coordinates; this work was only publicized outside Sweden in the early 1970s (Pred, 1973). As interpreted to a wider audience (Carlstein *et al.*, 1978), time and space are viewed as resources which constrain activity. Any behaviour which requires movement involves the individual or group traversing a path through space and time, as depicted in Figure 5.1 in which movements along the horizontal axis indicate spatial traverses while those along the vertical mark the passage of time. Any journey, termed a lifeline by Hägerstrand, thus involves movement along both axes simultaneously.

Constraints to movement in time and space operate in three ways. First, there are capability constraints restricting movement; in time, these refer to the biological need for about eight hours of sleep in every twenty-four, whereas the extent of movement in space during a given time is controlled by the available means of transport. Secondly, there are coupling constraints, which operate to require certain individuals and groups to be in particular places at stated times (teachers and pupils in schools, for example). And finally, authority constraints preclude an individual from being in certain places at set times. Together, these three define the time-space prism (Figure 5.1) which contains all the available lifeline paths for an individual starting at a certain location and needing to return there at a given time.

Pred (1977) has claimed of Hägerstrand's time-space geography that it

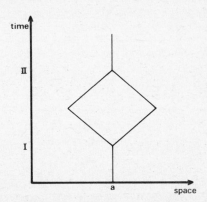

Figure 5.1   The time-space prism. In this simple example a person starts at point a: he cannot leave this point until time I and he must return to it by time II: the prism between those times indicates his maximum available spatial range.

has the potential for shedding new light on some of the very different kinds of questions customarily posed by 'old-fashioned' regional and historical geographers, as well as 'modern' human geographers . . . [it] holds the promise of identifying new questions of social and scholarly significance. It may also satisfy the longing shared by many for a more humanistically oriented geographic approach to some of the more complex and frustrating problems of modern society . . . it has the potential for enabling geographers to attain a new level of intellectual maturity. That is, through Hägerstrand's conceptual structure geographers may grow so bold as to seriously ask themselves what insights and assistance they can provide to researchers in the neighbouring social and biological sciences (p. 207).

Hägerstrand's intention with his framework is to provide a focus on the quality-of-life implications of the packing of people in time and space, and instead of simply looking, as in traditional geographical investigations, at parts of a structure it offers a contextual approach which views individuals' situations relative to other individuals within their environments. A new regional geography may emerge emphasizing 'principles of togetherness' (Pred, 1977, p. 213) rather than landscape components, and planning applications have been developed for the siting of facilities so that they are accessible within the constraints of time-space prisms (Carlstein, Parkes and Thrift, 1978; Pred and Palm, 1978). Thus, according to Pred (1977) although

> Time-geography is not a panacea for human geographers . . . It is, however, a great challenge . . . to cease taking distance itself so seriously . . . to accept that space and time are universally and inseparably wed to one another; to realize that questions pertaining to human organization of the earth's surface, human ecology, and landscape evolution cannot divorce the finitudes of space and time. . . . It is a challenge to turn to the 'choreography' of individual and collective existence − to reject the excesses of inter- and intra-disciplinary specialization for a concern with collateral processes (p. 218).

Hagerstrand's initial work, and its interpretation by a number of followers, is clearly set in the behaviouralist tradition outlined here; much inductive research has been undertaken seeking to identify the parameters of individual lifelines, or 'time-space biographies'. But Hägerstrand's intentions were much wider, and spill over into the humanistic concerns outlined later in this chapter. (Carlstein, 1980, p. 57, in addition, notes that Hägerstrand and his co-workers reject behaviourist 'Simple recording of what people did for how long, etc.', arguing that this could not lead to explanation. He argues, instead, for a structuralist approach in which the allocation of time and space resources is a constraint on human agency imposed by the forces of production: see also Chapter 6 and Giddens, 1981.) As van Paassen (1981) expresses it, Hägerstrand's work is essentially humanistic, aimed 'to provide insight into what is specifically human in man's nature, and . . . [to] elucidate the specific human situation' (p. 18). The central aspect of the individual's biography, according to this view, is 'corporealization' – territorial coexistence in time and space. Within a society, one must act as a neighbour:

> To succeed as a neighbor at the same time requires that one should act as a neighbor. This permits of a variety of action, sometimes in standard situations with well-known neighbors with recognizable, predictable behavior, sometimes in critical and/or unexpected situations (p. 25)

Hägerstrand seeks to investigate this neighbourliness, to understand it and to improve it.

Although Hägerstrand's initial work in time-space geography, following his earlier research on diffusion and migration, was interpreted as behaviourist, in effect it – and that which it stimulated – spreads over several of the approaches to human geography outlined in this book (see also p. 155 and 198). Within the behaviourist approach, it has encouraged joint consideration of space and time as impediments to behaviour. Thus Parkes and Thrift (1980, p. xi) have sought 'to place time firmly in the minds of human geographers', in a presentation that they term chronogeographical. This extends well beyond the work of Hägerstrand's group at Lund on biographies, developing the theme that

> Together, territories, societies and times are the principal components of urban and social geography (p. 34).

Methods in behavioural geography

Whereas the theory- and model-builders of the spatial-science school of human geographers received much of their stimulus from neo-classical economics, in some cases via regional science, the alliance for behavioural geography was largely with the social sciences having a major empirical content, notably psychology and sociology. The behaviouristic approach is an inductive one, with the aim being to build general statements out of observations of ongoing processes. The areas studied were very much determined by the work in the spatial science school. As Brookfield's statement quoted at the beginning of this chapter suggests, the models and theories of the latter, such as central place theory, raised many of the queries which the behaviourists, stimulated by their observations of the failings of such theories when matched against the 'real world', sought to follow up. In terms of the accepted route to scientific explanation (Figure 3.1), therefore, behavioural geography involved moving outside the accepted cyclic procedure to input new sets of observations on which superior theories might be based. In doing this, the behaviourists did not really move far from the spatial-science ethos. Indeed, many of their methods were those of their predecessors; Gould's mental map studies, for example, used the same technical apparatus as the factorial ecologies (p. 84).

Sowewhat away from this general orientation of behavioural work, Pred (1967, 1969) presented an ambitious alternative to theory-building based on 'economic men' in his two-volume work *Behavior and Location*. He began with a critique of existing location theory comprising three groups of objections: those concerned with logical inconsistency – it is impossible for competing decision-makers to arrive at optimal location decisions simultaneously; those concerned with motives – maximizing versus satisficing; and those concerned with human ability to collect, assimilate and use all possible information. Thus (Pred, 1967):

> Bunge's theoretical geography is easily distinguished from geographical location theory because its optimal final goals are disassociated from the interpretation of real-world economic phenomena . . . and because these same goals can only yield a body of theory that for all intents and purposes is totally abehavioral and static rather than dynamic (p. 17).

Decisions on locations and on land use are made with imperfect knowledge by fallible individuals. As a result there is bound to be some disorder in the ensuing spatial patterns.

As an alternative to the forms of theory-building which he attacked, Pred proposed the use of a behavioural matrix (Figure 5.2), whose axes are quantity and quality of information available and the ability to use that information: economic man is located in the bottom right-hand corner. Because of the nature and importance of information flows, position on the first axis depends in part on the decision-maker's spatial location; on the second, his position would reflect aspiration levels, experience, and the norms of any groups to which he belonged. Different people in different positions in the matrix would vary in their decisions, therefore; indeed, two at the same position may act on different bases and in different ways.

Individuals do not stay at the same position in the matrix, and spatial patterns are not static. Thus in his second volume, Pred (1969) introduced a dynamic element by shifting individuals through the matrix; as they shift, and change their decisions, so the environment changes for others. As people learn, they both acquire more and better information, and become more skilled in its use; they shift towards the bottom right-hand corner of the matrix, some of them in advance of others, who benefit from the activities of the 'decision-leaders'. The unsuccessful are gradually eliminated, so that with time a concentration of 'good' decision-makers close to the economic-man position evolves. But changes in the external environment produce parametric shocks, which result in decision-makers becoming less informed and less certain; as a consequence they are shifted back towards the upper left-hand corner and another learning cycle begins. As long as parametric shocks occur more frequently than the learning experience takes, an optimal location pattern will never emerge, except perhaps by chance.

Pred (1969) presented the behavioural matrix as a 'gross first approximation' (p. 9.1), arguing that any theory is better than none (p. 139), even if the model itself is literally untestable (p. 141: see also Pred and Kibel, 1968). Harvey (1969b) treated it to a scathing review, however, calling the two dimensions of the matrix vaguely defined, ambiguous, unoperational,

Figure 5.2   The behavioural matrix. Source: Pred (1967, p. 25).

and an oversimplification of the complex nature of behaviour. Indeed, Harvey (1969c) was generally sceptical about the potential of a behavioural location theory, a view shared by Olsson (1969), who pointed out the difficulties of studying processes and demonstrated that much behavioural geography involved only the inferring of processes from aggregated data on individual behaviour. Others argued that such inference could be very strong, however. Rushton (1969), for example, accepted that any one pattern of behaviour – what he termed *behaviour in space* – was largely a function of the spatial structure within which it occurred (the choice of shopping centres by a population, for example), but he claimed that general rules of *spatial behaviour* could be deduced from examining the types of preferences displayed within a particular pattern:

> To say that these preferences do not exist independently of the environment where the decision is made is to argue that environments could exist about which the person would be unable to reach a decision (p. 393).

Thus a distance-decay pattern reflects the details of the environment in which it is observed, but to produce it people must make decisions based on certain rules which they apply, and Rushton developed a procedure for isolating those rules. Its validity has been queried, however (Pirie, 1976), and, like Pred's behavioural matrix, it has not attracted many disciples. (Rushton's arguments suggest activity in a 'taken-for-granted world: see p. 155.)

Harvey (1969c) suggested two alternatives to behavioural location theory – further development of normative theory and the construction of a stochastic location theory. He felt that both of these offered more immediate pay-offs in terms of understanding spatial patterns than did behavioural efforts, because of the conceptual and measurement problems of the latter.

The stochastic approach was also favoured by Curry (1967b), who argued that large-scale patterns are the outcomes of small-scale indeterminacy; individual choices may be random within certain constraints, but when very many of them are aggregated they may display considerable order (see above, p. 120). Similarly, Webber (1972) attempted to model locational decision-making processes in states of uncertainty using normative approaches: his conclusion was that 'uncertainty increases agglomeration economies' (p. 279), thereby leading to greater concentration of economic activities and people into cities than would be predicted by models based on 'economic men'. Game theory is a mathematical procedure developed to handle decision-making in uncertainty, but it has received little attention from human geographers (Gould, 1963). The most ambitious attempt to build on such theories, which has had very little impact, is Isard *et al.* (1969).

The behavioural approach has not brought about a revolution away from the spatial-science paradigm, therefore, and in effect has become an attachment to it. Whereas various normative approaches start with certain assumptions, usually simplifying ones, about human behaviour, and then deduce what spatial patterns follow from such axioms, the behaviourists have sought to modify the assumptions by an inductive procedure which seeks the rules of behaviour that can be used to predict (and therefore explain) spatial patterns (Gale and Golledge, 1982). In its entirety, this latter approach involves a sequence of interrelated investigations. An individual is faced with a decision, either one with a direct spatial input or one with spatial consequences. To make that decision, he sets criteria, collects information, and then evaluates the information against the criteria. As a result of the evaluation, a decision to act might be made, or instead he determines to change his criteria, to collect more information, or to do both. (In this way, his problem-solving is very similar to the procedures of the scientific method outlined in Figure 3.1.) Many investigations have not gone through this full sequence, however, but have focused on certain aspects of it, such as the flow of information, from which characteristics of other elements in the sequence may be inferred.

Whereas very many of the studies in the spatial science school reviewed in earlier chapters could be conducted using either published data sources (such as censuses) or relatively small field-collection exercises, the behavioural approach has required much more effort on the part of human geographers in data collection from the individual decision-makers. This need to conduct social surveys of various kinds has furthered the growing links between geographers and sociologists, psychologists, and, to a lesser extent, political scientists, and has led to an expansion in the data-handling procedures necessary for the training of geographers. One of the problems in their use is that many of the topics studied in human geography involve very large numbers of individual decisions – as in migrations, journeys to work and to shop, voting decisions, and so on. Very large sample surveys may be necessary to produce valid generalizations about such behaviour, but resource limitations have meant concentration on both small selections and only limited segments of the full behavioural sequence. Thus, for example, Brown and Moore's (1970) schema for the study of intra-urban migration decisions has mainly been tested in part only (e.g. Clark, 1975, 1981). More success with use of the behavioural sequence has probably been achieved in those branches of human geography that deal with topics involving relatively few decision-makers. In the study of diffusion, for example, Brown (1975) has attempted to divert attention away from overall patterns of spatial spread and the reasons for adoption (or not) to the decision-making which brings certain innovations to places; most of these innovations involve the selling of products, and so he terms them 'consumer innovations' (see also Brown, 1981). Similarly, a number of industrial geographers have moved from investigations of aggregate patterns, which could be compared to those predicted by application of neo-classical economic analysis (Smith, 1971), to the study of decision-making behaviour by firms (e.g. Hamilton, 1974; Hayter and Watts, 1983).

Behavioural geography is now widely accepted within the positivist orientation. It seeks to account for spatial patterns within the environment (both man-made and natural) by establishing generalizations about man-environment interrelationships, and then using these as a basis for change through environmental planning activities that 'modify the stimuli which affect the spatial behavior of ourselves and others' (Porteous, 1977, p. 12). According to Gold's (1980) text, this approach is built on four main features:
1 The environment in which individuals act is that which they perceive, which 'may well differ markedly from the true nature of the real world' (p. 4).
2 Individuals interact with their environments, responding to them and reshaping them.
3 The focus of study is the individual, not the group.
4 Behavioural geography is multidisciplinary.
Within this general framework, research methods vary substantially but the general orientation – inductive generalization leading to planning for environmental change – remains. Eventually, it is hoped, a 'powerful new theory' (Cox and Golledge, 1969) will emerge. As yet, it has not (Golledge, 1981a, p. 1338; see also p. 132), but Golledge argues that substantial advances in understanding spatial behaviour have been made by studying 'individual preferences, opinions, attitudes, cognitions, cognitive maps, perceptions, and so on' (p. 1339) – what he terms process variables.

Two main types of work have been identified within this approach (see Clark, 1981; Pipkin, 1981; Thrift, 1981). The first is based on the study of overt behaviour and uses the traditional positivist formulation of dependent variable (choice of travel mode perhaps) influenced by independent variables (characteristics of the decision-maker). Developments in this have involved extending the range of statistical techniques applied by geographers (Burnett, 1981; Wrigley, 1979). The other approach involves attempts to identify the mental constructs that lie behind the overt behaviour. As Pipkin notes, although considerable effort has been expended on the identification, little has yet been achieved in linking cognitive schemata to behaviour, and thus extending the predictive models of the first approach.

Cultural and historical behavioural geography

As indicated in earlier chapters, workers in cultural and historical geography were neither closely involved in nor attracted by the so-called quantitative and theoretical revolutions. There was some application of statistical procedures, as attempts to make work in these fields appear more 'modern' – several of these attempts were by non-historical (e.g. Pitts, 1965) and non-cultural geographers – but few changes of substance in the approaches adopted. But by the 1970s, historical and cultural geographers had taken the initiative, and were proposing alternative philosophies to that of positivism, philosophies which were humanistic in their orientation.

The beginnings of these changes can be traced to two papers, one of which had more impact than the other on the geographical discipline at large. In the first, John K. Wright (1947) introduced the term *geosophy*, defined as the study of geographical knowledge:

> it covers the geographical ideas, both true and false, of all manners of people – not only geographers, but farmers and fishermen, business executives and poets, novelists and painters, Bedouins and Hottentots – and for this reason it necessarily has to do with subjective conceptions (p. 12).

Wright conceded that study of such subjective ideas was not open to the employment of the strict scientific principles of physical geography, but claimed that it provided indispensable background and perspective to geographical work:

> geographical knowledge of one kind or another is universal among men, and is in no sense a monopoly of geographers . . . such knowledge is acquired in the first instance through observations of many kinds. . . . Its acquisition, in turn, is conditioned by the complex interplay of cultural and psychological factors . . . nearly every important activity in which man engages, from hoeing in a field or writing a book or conducting a business to spreading a gospel or waging a war, is to some extent affected by the geographical knowledge at his disposal (pp. 13–14).

These words could well have heralded an earlier start to behavioural geography than that chronicled earlier in this chapter, but apparently, by the lack of reference to them in later published works, they had little impact until taken up by Wright's colleague at the American Geographical Society, David Lowenthal (1961), in a widely cited paper 'concerned with *all* geographic thought, scientific and other: how it is acquired, transmitted, altered and integrated into conceptual systems' (p. 259). In a wide-ranging survey, Lowenthal argues that the world of each individual's experience is intensely parochial and covers but a small fraction of the total available. There are consensus views about many aspects of the world, but often an individual will mistakenly assume that his view is the consensus. We all live in personal worlds, which are 'both more and less inclusive than the common realm' (p. 248). Our perceptions of these personal worlds are personal too; they are not fantasies, being firmly rooted in reality, but because 'we elect to see certain aspects of the world and to avoid others' (p. 251) behaviour based on such perceptions must have its unique elements. Different cultures have their own shared stereotypes, however, which are often reflected in language, and they attempt to create environments fitting into these stereotypes:

> The surface of the earth is shaped for each person by refraction through cultural and personal lenses of custom and fancy. We are all artists and landscape architects, creating order and organizing space, time, and causality in accordance with our apperceptions and predilections (p. 260).

These ideas were put into practice in other papers concerned with the interpretation of landscapes as reflections of societal norms and tastes (e.g. Lowenthal, 1968; Lowenthal and Prince, 1965), thereby belatedly bringing Wright's ideas before a wider, and more readily appreciative perhaps, audience.

The second of the original papers was by a British geographer, although it was published in India (Kirk, 1951): the main arguments were reiterated in a later article (Kirk, 1963). He, too, stressed that the environment is not simply a 'thing' but rather a whole with 'shape, cohesiveness and meaning added to it by the act of human perception' (Kirk, 1963, p. 365): once this meaning has been ascribed, it tends to be passed to later generations. Thus Kirk recognizes two, separate but not independent, environments; a phenomenal environment, which is the totality of the earth's surface, and a behavioural environment, which is the perceived and interpreted portion of the phenomenal environment:

> Facts which exist in the Phenomenal Environment but do not enter the Behavioural Environment of a society have no relevance to rational, spatial behaviour and consequently do not enter into problems of the Geographical Environment (p. 367).

Since much of geography is concerned with decision-making and its consequences, appreciation of the behavioural environment should be central to the study of geography. Indeed, according to Ley (1977a), one cannot proceed without such awareness of what is in the behavioural environment; even an apparently neutral statement such as 'Pittsburgh is a steel town' is, he argues, a value-laden view of a geographer-outsider, which may not accord with the perceptions of the resident-insiders. Thus:

> Too often there is the danger that our geography reflects our own concerns, and

not the meanings of the people and places we write of. . . . The geographical fact is as thoroughly a social product as the landscape to which it is attached (p. 12).

. Scientific disciplines, as indicated in Chapter 1, can be divided into their 'invisible colleges', groups of scholars working on the same topic who refer to each other's publications. Wright, Lowenthal and Kirk were not members of any major college during the early 1960s, and so had very little impact on the first phase of behavioural work identified earlier in this chapter. (None of the three, for example, is in Pred's (1967, 1969) bibliographies, nor is any one referred to by Wolpert: Golledge, Brown and Williamson (1972, p. 75) make only a passing reference to Lowenthal's work–'Pursued by insightful researchers, the analysis of literary and other artistic data of past and present can have strong explanatory power. The subjective element in these attempts to assess the impact of spatial perception is acknowledged, but its presence in many other studies is more subtle and potentially damaging'. Note, however, Kirk's, 1978, p. 388, reference to his influence on non-geographers who adopted his concept of the behavioural environment.) The behavioural work of the mid-1960s was in the positivist mould; that of the three authors just discussed was not.

Several cultural and historical geographers took up the concepts of the behavioural or perceived environments. Among the former, one of the leaders was Brookfield, a British geographer with field experience, by the mid-1960s, of South Africa, Mauritius, New Guinea, and the Pacific Islands. In reviewing work by cultural geographers on alien societies, he noted (Brookfield, 1964):

> A difference of approach is apparent between those who have an overtly chorographic purpose, who scarcely ever seek explanations in matters such as human behaviour, attitudes and beliefs, social organization, and the characteristics and interrelationships of human groups, and those whose inquiries are not primarily chorographic, and who are more inclined to undertake a search for processes as a means of reaching explanation (p. 283).

Social organization, Brookfield argued, is the key to many explanations, so that:

> when an individual human geographer is sitting down in one small corner of a foreign land, and seeks to interpret the geography of that small corner, then it is difficult for him to do so without trying to comprehend the perception of environment among the inhabitants (p. 287).

But geographers had largely failed to delve into such details of social organization, because of the broad areal scale at which they had tended to work, their concern with distributions rather than with processes, and their avoidance of what he terms 'micro-geography'. Inquiry in human geography should involve three stages:

1   general statements about areal patterns and interrelationships;

2   detailed local inquiries which follow up the questions about processes raised by these general statements; and

3   organization of the general and local material to produce explanatory generalizations.

Brookfield's argument was for more micro-geographical studies at the second of these stages, providing the basis for the development of comparative methods with which generalizations could be forged (Brookfield, 1962).

Brookfield (1969) later surveyed the literature which showed that 'decision-makers operating on an environment base their decisions on the environment as they perceive it, not as it is. The action resulting from decision, on the other hand, is played out in a real environment' (p. 53). Referring to the 'modern' behavioural work, as well as the studies of cultural geographers, he pointed out the great problems involved in isolating the perceived environment – something which is 'complex, monistic, distorted and discontinuous, unstable and full of unwoven irrelevancies' (p. 74) – and in building it into an analytical methodology. Further data are needed, too, on such topics as work-organization, time-allocation and budget-allocation, on the meaning of consumption and of distance – all necessary tasks for the full understanding of man-environment systems.

The concept of the perceived environment has a considerable pedigree in historical-geographical scholarship, though without the current terminology; a major example of its use is Glacken's (1956, 1967) survey of societal attitudes to environments. Historical geographers, it has been suggested (Prince, 1971a), must study a trilogy of worlds: the real world, as recorded in documents and in the landscape; the abstract world, as depicted by general models of spatial order in the past; and the perceived world: 'Past worlds, seen through the eyes of contemporaries, perceived according to their culturally acquired preferences and prejudices, shaped in the images of their assumed worlds' (p. 4). From these three, it may be possible to provide explanations of landscape changes, which cannot be obtained from the assumed processes provided by continua of data over time (see also Moodie and Lehr, 1976). Reconstruction of past environments is extremely difficult, however, for it involves seeing the written record through the cultural lens of the writer:

> A study of past behavioural environments provides a key to understanding past actions, explaining why changes were made in the landscape. We must understand man and his cultures before we can understand landscapes; we must understand what limits of physical and mental strain his body will bear; we must learn what choices his culture makes available to him and what sanctions his fellows impose upon him to deter him from transgressing and to encourage him to conform (Prince, 1971a, p. 44).

Perhaps the enormity of such a task is the reason why most success in such reconstructions has been with regard to the relatively recent past, such as of the perceptions which guided the settlement of the American West (e.g.

Lewis, 1966), although Wright (1925) essayed a similar task for Europe at the time of the Crusades.

Not all perceived worlds refer to either past or present; some landscapes have been fashioned by men out of their Utopian views of the future (Porter and Lukermann, 1975; Powell, 1971). In general, however, and whether of past, present or future, geosophy has not become a popular field of study, despite some intriguing essays (see Lowenthal and Bowden, 1975). But directly or indirectly, it has led to arguments for alternative approaches in human geography to those of the positivist, and it is those arguments which form the material for the rest of this chapter.

The attack on positivism and the humanistic approaches

From the early 1970s, work by cultural and historical geographers has been presented as an attack on the positivism of spatial science. To replace the latter, a variety of humanistic approaches has been proposed, presenting a focus on the decision-maker and his perceived world and denying the existence of an objective world which can be studied by positivist methods. The intent is to reorient human geography towards a more humanistic stance, to resurrect its synthetic character, and to re-emphasize the importance of studying unique events rather than the spuriously general.

Anti-positivism, idealism, and historical geography

The various humanistic approaches have much in common, but they can be separated into different proposals in the present context. The first to be discussed here is associated with two workers who were together at the University of Toronto at the end of the 1960s; both are historical geographers.

The basic theme of a first paper (Harris, 1971) is that geography is a synthetic discipline, concerned with particular assemblages of phenomena and not with the science of spatial relations. Thus:

> When the history of North American geography in the 1950s and 1960s is written, a paradox with which it will have to deal is how, with little argued, logical justification, so many geographers came to see their subject as a science of spatial relations (p. 157).

With May and Sack (p. 107), Harris sees the spatial perspective leading to the dismemberment of geography, as specialists communicate more with their contemporaries in other disciplines than with geographers, and develop theories which are descriptions of how the world might operate under certain conditions, rather than of how it does actually work.

> The difficulty in conceiving of geographical theory comes down to this. The development of theory is necessarily an exercise in abstraction and simplification in which the complexities of particular situations are eliminated to the point which common characteristics become apparent. But if geography is thought to have a particular subject matter, it is certainly not individual phenomena or

categories of phenomena which other fields do not study. Rather it is a whole complex of phenomena, many or all of which may be studied individually by other fields but which are not studied elsewhere in their complex interactions (p. 162).

The clear parallel, according to Harris, is with history, for:

> Few historians would attempt to develop a general theory of revolutions. In so doing they would lose grasp of the type of insight that characterizes good historical synthesis (p. 163).

The goal of both history and geography is synthesis, therefore. In developing syntheses, positivist methods may be applicable. Historians may be law-consumers, applying the generalizations of other social scientists to particular events, and geographers could operate likewise; alternatively, both historians and geographers could apply the idealist method, arguing that all activity is based on personal theories. Thus Harris (1978) writes elsewhere of a 'historical mind':

> Such a mind is contextual, not law-finding. Sometimes it is thought of as law-applying but, characteristically, the historical mind is dubious that there are overarching laws to explain the general patterns of human life (p. 126).

This mind, he argues, is open and eclectic, uses no formal research procedures, sees things in context, is sensitive to motives and values, excludes little, and is wary of sweeping generalizations. Its goal is understanding, not planning, and this should be the case with the 'geographical mind' too. (To understand an event is to appreciate why it took place, and is the humanistic goal: to explain an event is to predict it, as an instance of a general law or suite of such laws, and is the positivist goal.)

Developing his theme of the parallel between history and geography, Harris (1971, p. 167) suggests four major points of agreement concerning the nature of history:

1 Its primary concern is with the particular;
2 Explanation may take into account the thoughts of relevant individuals;
3 Explanation may make use of general laws; and
4 'Explanation in history relies heavily on the reflective judgement of individual historians'.

From these, he then argues:

> If geography aims to describe and explain not so much particular events or peoples, as particular parts of the surface of the earth, then these points of agreement about history also apply to geography (p. 167).

(The term 'particular parts' can be widely interpreted, it would seem, for Harris (1977) has himself sought to understand the nature of north-western European colonizing societies 'by a model': the implication is that colonists developed a common reaction to the 'new world'.) The landscape results from actions; behind those actions lie thoughts; study of thoughts allows understanding of landscape. Thus synthesis is crucial, since:

the idea of synthesis itself becomes more important as it becomes obvious that our larger problems transcend narrow subject-matter fields . . . integration . . . in a larger understanding is still achieved, however aided by statistical methods and computers, by the judgement of wise men who have cultivated the habit of seeing things together (p. 170).

Geographers, presumably, are to be those wise men; not an original claim for, according to Buttimer (1978a), the basis of Paul Vidal de la Blache's work was that:

The task which no other discipline with the possible exception of history claims is to examine how diverse phenomena and forces interweave and connect with the finite horizons of particular settings. Temporality and spatiality are universal features of life so historical and geographical study belong together (p. 73).

Positivist work seeks the same end – interweaving parts of a whole – but its parts are instances of general laws, not unique events.

Many of Harris's arguments were extended by his Toronto colleague, Leonard Guelke, whose first paper (Guelke, 1971) was a strong criticism of the 'narrowly conceived scientific approach' (p. 38) to geography using the positivist method. Thus he argued against geography as a law-seeking activity by asking the proponents of the positivist approach to indicate how their laws would meet the basic standards of scientific acceptability, particularly with regard to prediction. Whereas they might be able to produce generalizations concerning the phenomena which they actually studied he felt it very unlikely that they could define laws applicable to all examples of the relevant phenomena. Statistical regularities are not laws and

Until the new geographers have shown that the laws that might conceivably be discovered in geography will be more than generalizations, which describe common but non-essential connections between phenomena, their claims must be treated cautiously . . . there is little cause for optimism, especially as the statistical methods widely employed by geographers cannot be considered appropriate law-finding procedures (p. 42).

Regarding geography as a law-applying science, Guelke argues that laws of human behaviour are virtually impossible to conceive in anything but the most generalized form, because so much behaviour is culturally–specific, and an a priori statement of the determining conditions for their operation is not feasible. Thus 'Human geographers cannot consider themselves to be law-applying scientists . . . because they have no laws to apply' (p. 45).

Turning to the use of theories and models in geography, Guelke points out that for them to serve a valid purpose in the pursuit of understanding, criteria must be erected which indicate how such devices are testable against reality (see also Newman, 1973, on the vague use of the term hypothesis). Such criteria have not been, and cannot be, stated, Guelke claims; to him, studies purporting to test central place theory seem to operate the rule that 'one counts one's hits but not one's misses' (p. 48; see also Guelke, 1978, p. 50). Too often, failure to reproduce reality is explained by claims that the test environment was not entirely suitable, and frequently *ad hoc* hypotheses are adduced to account for observed disparities. Models and theories may have heuristic value for human geographers, clarifying certain aspects, therefore, but they can have no explanatory power.

Guelke's (1971) conclusions are that:

> The new geography . . . has not yet produced any scientific laws and . . . appears unlikely to produce them in the future. . . . The theories and models . . . are not amenable to empirical testing. . . . The new geographers insisted on . . . logical and internally consistent theories and models. Yet, none of their theoretical constructs were ever complex enough to describe the real world accurately. They had achieved internal consistency while losing their grip on reality (pp. 50–1).

His alternative to the so discredited procedure is the idealist approach mentioned by Harris, which is (Guelke, 1974) 'a method by which one can rethink the thoughts of those whose actions he seeks to explain' (p. 193). All actions, according to the idealist, are the result of rational thought, the parameters of which are constrained by a theory, which in turn is 'any system of ideas that man has invented, imposed, or elicited from the raw data of sensation that make connections between the phenomena of the external world' (p. 194). Many such theories are part of the society and culture inhabited by the actor under consideration, and include its religions, myths and traditions. Using them, 'the explanation of an action is complete when the agent's goal and theoretical understanding of his situation have been discovered. . . . One must discover what he believed, not why he believed it' (p. 197; but see Sayer, 1981). Thus human geographers do not need to develop theories, since the relevant theories, which led to the action being studied, already exist (or existed) in the minds of the actors. The task of the analyst is to isolate those theories. Some of them may be unique to paticular individuals – such as that which led Columbus to sail westwards – but there are many consensus theories, shared in large part by large numbers of actors; they represent the order which man has himself stamped on the world, and do not require further theories in order to be understood.

Guelke's argument was challenged by Chappell (1975), who pointed out that by focusing on the individual actor alone the idealist omitted any reference to the environmental constraints and influences on his actions (see also Gregory, 1978a). Guelke (1976) accepted the existence of such constraints and influences, but claimed that investigation of them lay outside the geographer's domain. Study of environmental causes would, he felt, lead into physiology and psychology and deviate attention from 'the most critical dimension of human behaviour, namely the thought behind it' (p. 169). Chappell (1976) responded that 'to go so far as to say that there is no possible respectable theory to explain man's rational theories and the actions which flow from them' (p. 170) is to be myopic: 'paradigms not only explain facts but they guide the research of whole disciplines' (p. 171). To him, Guelke's contention that the ultimate causes of man's actions lie outside the scope of human geography places geographers in an inferior position in the academic division of labour.

In a further essay, Guelke (1975) addressed his ideas on idealist approaches to historical geographers, as a counter to the arguments that they should adopt the approaches and techniques of positivism (see p. 89). He argued:

> It is obvious that quantitative techniques will often be useful. . . . Statistical methods put in harness with positivist philosophy are a dangerous combination. . . . Historical geographers need to rethink not their techniques but their philosophy. . . . This can best be achieved by moving from problem-solving contemporary applied geography towards the idealist approach widely adopted by historians. (p. 138).

Gregory (1976) agreed with the first part of this statement, but not with the proposed solution. Like Chappell, he saw the need to investigate individual action within its constraining structures (see below, p. 181).

Positivist approaches in human geography have been defended against the idealist attack by Hay (1979a), who both responds to the criticisms and in return raises points of contention in the proposed alternative. Thus he argues that Guelke's case is ill-founded and rests on misconceptions of the nature of positivism, such as that all theory must be both normative and based on conceptions of economic man, that to be scientific is to be nomological, and that prediction is the same as prophesy (rather than simply testing, from the known to the unknown). Further, he claims that Guelke presents an anti-positivist argument by using a positivist test, and that he fails to realize the value of *ad hoc* hypotheses in the improvement of theory (which is the basis of Lakatos's concept of a research programme and its positive heuristic – p. 24): Guelke should not, according to Hay, ask 'does this theory explain Y?' but rather should ask 'does this theory contribute to an understanding of Y?'

With regard to the idealist alternative, Hay raises the problem of studying groups rather than individuals. To Guelke (1978):

> The assumption that thought lies behind human action is not related to the numbers involved. . . . If thousands of people drive motor cars to their places of work the idealist assumes that each of these journeys is a considered action involving thought. In such situations the investigator will not be able to look at each case individually, but he will seek to isolate the general factors involved in typical circumstances . . . [for which he] might well make use of statistical procedures . . . the value of statistical analysis will largely depend on its successful integration in the general interpretation or explanatory thesis being developed (p. 55),

which is a procedure akin to that employed by the behaviourist geographers whose work has been reviewed earlier in this chapter. Such a procedure does not give ontological status to groups as collections of individuals in which, as with 'traditional' regions, the whole is greater than the sum of the parts.

Secondly, Hay points out the objective facts must influence behavioural outcomes, in addition to the thoughts of the actors: Columbus found America, because it was there. Thirdly, he claims that the idealist position ignores the possibility of either unconscious or subconscious behaviour. In sum, idealism is reductionist, but the world is more than a large number of independent decision-makers.

Idealism has also been criticized by Mabogunje (1977) who claims that 'Such a retreat from objective theory formulations as a means of seeking explanation to certain events would exclude from our consideration the exploration of the consequences of societal actions' (p. 368) and that instead of retreating to a focus of particular cases – 'seek[ing] special explanation for each situation in which a different value system can be shown to be operative' (p. 370) – geographers should attempt to build better theories encompassing these differences in value-systems. Others have asked how an idealist interpretation can be verified. (See also the same question regarding phenomenology, raised by Mercer and Powell (1972), who wonder whether two phenomenologists can ever have the same 'intuitions' of a phenomenon, or indeed know whether they have.) Guelke is prepared for this argument, presenting the analogy of the court-of-law, rather than the positivist's laboratory, for his Popperian procedure.

A well-verified idealist explanation will be one in which a 'pattern of behaviour can be shown to be consistent with certain underlying ideas. Where data are presented which are not in accordance with a proposed explanation a new hypothesis will be needed' (Guelke, 1978, p. 55), which is somewhat in contradiction to his earlier statements that theory should not be imposed from without by the observer (p. 147). Even so, he says, 'one cannot guarantee mistake-free interpretations. The complex nature of human societies and lack of pertinent data makes it inevitable that many idealist interpretations will be of a tentative character' (p. 55), a position which is very similar to Moss's (1977) outline of deductive procedures in historical explanation.

The idealist philosophy, according to Guelke (1981), combines two positions: a metaphysical argument that 'mental activity has a life of its own which is not controlled by material things and processes'; and an epistemological argument

> that the world can only be known indirectly through ideas . . . all knowledge is ultimately based on an individual's subjective experience of the world, and comprises mental constructs and ideas. There is no 'real' world that can be known independently of mind (p. 133).

Positivist spatial science is thus criticized because it believes in the existence of a 'real' world, the nature of which it seeks to explain via general laws of behaviour. Behavioural geography is similarly criticized, not for its acceptable (to Guelke) premise that behaviour can be understood as a response to perceived images and subjective evaluations, but because of two assumptions within this field: 'that identifiable environmental images exist that can be measured accurately . . . [and] that there are strong relationships between revealed images and preferences and actual (real-world) behaviour' (Bunting and Guelke, p. 453). Such an approach, Bunting and Guelke believe, traps human geographers into a single-cause model, much like the earlier environmental determinism (see p. 41), and even if one accepted this model research in behavioural geography has failed to validate it. They argue for an idealist perspective, which focuses on overt behaviour and its interpretation: 'In searching for the truth a scholar conducts a critical dialogue with his evidence and in due course he puts the results before his colleagues for their appraisal' (p. 458). In response, Downs (1979), Rushton (1979), and Saarinen (1979) claim that Bunting and Guelke misrepresent much of behavioural geography and argue that although the study of overt behaviour may give clues as to the answers to 'why?', only the study of decision-making can provide an understanding that is of value in planning, as in the work on environmental hazards. Golledge (1981a, p. 1328) also responding to Bunting and Guelke, claims that to qualify as behavioural geography research may focus on the overt act, but more importantly the set of explanatory variables must include 'one or more process variables' (see above, p. 139).

Phenomenology and related approaches

Phenomenology has attracted more attention among human geographers than has idealism. The term was first used by a geographer as long ago as 1925 (by Sauer) but it has only become widely known during the 1970s.

The first direct statement advocating a phenomenological approach was by Relph (1970), who has also been associated with the Department of Geography at the University of Toronto. Despite a variety of specific interpretations, he noted that the basic aim of phenomenology is to present an alternative methodology to the hypothesis-testing and theory-building of positivism, an alternative grounded in man's lived world of experience. Phenomenologists argue that there is no objective world independent of man's existence – 'all knowledge proceeds from the world of experience and cannot be independent of that world' (p. 193). Thus, according to Entrikin (1976), 'phenomenologists describe, rather than explain, in that explanation is viewed as [an observer's] construction and hence antithetical to the phenomenologist's attempt to "get back" to the meaning of the data of consciousness' (p. 617).

Phenomenology in human geography is concerned with what Kirk termed the phenomenal environment (p. 141). The contents of that environment are unique to every individual, for each of its elements is the result of an act of intentionality – it is given meaning by the individual, without which it does not exist but with which it influences behaviour. Phenomenology is the study of how such meanings are defined. It involves the researcher seeking to identify how the individual structures the environment in an entirely subjective way; the researcher is presuppositionless, using none of his own ideas in seeking to understand those of his subject. (Thus the subjectivity is that of the focus of the study, not that of the student who, in the positivist method, imposes his own subjective views of the world: Ley, 1980.) The phenomenologist may be satisfied with such empathetic understanding. Some seek to go further, however, to identify essences – elements in individual consciousness which control the allocation of meanings (Johnston, 1983b). Phenomenology, then, studies human appraisals. It works at the level of the individual, but may seek to identify the common (imprinted not agreed) elements of those appraisals, including those of scientists.

Relph's paper was followed by another from a geographer having associations with the University of Toronto, Yi-Fu Tuan (1971), to whom geography is the mirror of man, revealing the essence of human existence and striving: to know the world is to know oneself, just as careful analysis of a house reveals much about both the designer and the occupant. Thus the study of landscapes is the study of the essences in the societies which mould them, in just the same way that the study of literature and art reveals much of human life. Such study, by geographers, has a clear base in the humanities, rather than the social or physical sciences. Tuan (1974, 1975b) has illustrated it in a number of essays, giving, for example, insights into such topics as the sense of place.

> Humanistic geography achieves an understanding of the human world by studying people's relations with nature, their geographical behaviour as well as their feelings and ideas in regard to space and place (Tuan, 1976, p. 266). Scientific approaches to the study of man tend to minimize the role of human awareness and knowledge. Humanistic geography, by contrast, specifically tries to understand how geographical activities and phenomena reveal the quality of human awareness (p. 267).

Tuan exemplifies this with five themes: the nature of geographical knowledge and its role in human survival; the role of territory in human behaviour and the creation of place identities; the interrelationships between crowding and privacy, as mediated by culture; the role of knowledge as an influence on livelihood; and the influence of religion on human activity. Such concerns are best developed in historical and in regional geography; their value to human welfare is that they clarify the nature of the experience (see also Appleton, 1975). Indeed, Tuan (1978) claims that 'The model for the regional geographers of humanist leaning is . . . the Victorian novelist who strives to achieve a synthesis of the subjective and the objective' (p. 204). He quotes the first two pages of E.M. Forster's *A Passage to India* as a paradigm example.

Among others advocating the phenomenological approach, Mercer and Powell (1972) echo the arguments that application of positivism in geography 'left the subject with too many technicians and a dearth of scholars' (p. 28). Land-use patterns, they claimed, can never be understood 'by the elementary dictates of geometry and cash register' (p. 42); the world can only be comprehended through man's intentions and his attitudes towards it. In a lengthy discussion of the nature of phenomenology and its development in other disciplines, notably sociology, they point out 'a very real danger of the research worker assuming that concepts which are cognitively organized in his own mind "exist" and are equally clearly organized in the minds of his respondents' (p. 26), and argue instead for research methods which lead to empathy between observer and observed. Within geography this requires 'that we make every effort to view problems and situations not from our own perspective, but from the actor's frame of reference' (p. 48) – which is a scientific position of 'disciplined naivete'.

In similar vein, Anne Buttimer made a general case for study by geographers of the values which permeate all aspects of living and thinking (Buttimer, 1974). She feels that the order, precision and theory developed by positivist social sciences are dearly bought – 'we often lose in adequacy to deal with the values and meanings of the everyday world' (p. 3) – and that the behavioural geography of the sort discussed earlier in this chapter does not break away from the mechanistic, natural-science view of man (see also Ley, 1981, who points to the positivist elements in behavioural geography and the suggestion of operant conditioning: man reacts in predetermined ways to various stimuli, and can be manipulated accordingly. On the other hand 'An existentially aware geographer is . . . less interested in establishing intellectual control over man through preconceived analytical models than he is in encountering people and situations in an open, inter-subjective manner' (p. 24). The results of such activity are 'a meditation on life'; geographers would provide more comprehensive mirrors on life experience than is possible for their colleagues from more specialized disciplines, thereby clarifying the structural dynamics of life. Prediction would be impossible, apart for the 'most routinized aspects of experience' (p. 29), but the deeper understanding achieved would allow much more valid social action and planning than is currently possible.

In a later paper, Buttimer (1976) directs geographers' attention to the concept of the lifeworld, that amalgam of the worlds of facts and affairs with those of values which comprise personal experience – 'the pre-reflective, taken-for-granted dimensions of experience, the unquestioned meanings, and routinized determinants of behaviour' (p. 281). Positivism is rejected as a method for analysing the lifeworld because it separates the observer from that which he is studying; as a result, he fails to appreciate the human experience. Idealism is rejected, too, because it accepts that there is a real world outside the individual's consciousness. Phenomenology, on the other hand, is a path to understanding, on which informed planning can be built:

> It helps elucidate how . . . meanings in past experience can influence and shape the present . . . extremely important as preamble not only to scientific procedure, but also as a door to existential awareness. It could elicit a clearer grasp of

value issues surrounding one's normal way of life, and an appreciation of the kinds of education and socialization which might be appropriate for persons whose lives may weave through several milieux (p. 289).

The result is an understanding of man's actions as he understands them, rather than in the terms of abstract, outsider-imposed models and theories. And having achieved that understanding, the human geographer can transmit it to the subject, thereby helping him to understand himself and realise his potential. In this way, the applied geographer acts as a provocateur, stimulating human development but not forcing it (Buttimer, 1979).

Berry (1973b), too, has backed this phenomenological orientation, calling for

> a view of the world from the vantage of *process metageography*. By metageography is meant that part of geographic speculation dealing with the principles lying behind perceptions of reality, and transcending them, including such concepts as essence, cause and identity (p. 9).

But not all believe that phenomenology can entirely replace the positivist approach. Walmsley (1974), for example, accepts the merits of the case just presented because so many human decisions are based on 'experiential' rather than 'factual' concepts, but feels that the scale of geographical enquiry, and its long tradition of certain types of empirical work, will require maintaining the positivist orientation. That the perceived world is not necessarily the same as the real world will need to be realized, but 'logical consistency and empirical truth will remain central to geographical enquiry provided the importance of values is recognized' (p. 106).

Gregory (1978a) is critical of both positivism and phenomenology. Those favouring the former are criticized for making 'social science an activity performed *on* rather than *in* society, one which portrays society but which is at the same time estranged from it' (p. 51) and for supporting a procedure which, because it so often assumes *ceteris paribus* in testing its models, can never be sure why these fail, when they do, to replicate reality (p. 66). The necessity for humanistic approaches is recognized, but these will not be sufficient to provide a satisfactory foundation of themselves, because they ignore the 'constraints on social action which are so much part of the taken-for-granted life-world of the actors' (Gregory, 1978b, p. 166). Thus:

> A geography of the life-world must therefore determine the connections between social typifications of meaning and space-time rhythms of action and uncover the structures of intentionality which lie beneath them (Gregory, 1978a, p. 139).

But

> A major deficiency . . . is [the] restricted conception of social structure: in particular, it ignores the material imperatives and consequences of social actions and the external constraints which are imposed on and flow from them (*ibid*).

Phenomenology and idealism, then, must be incorporated with investigations of those imperatives and constraints; such incorporation produces a critical science, whose nature is discussed in the next chapter.

Whereas a key feature of the positivist/spatial-science approach has been a great numerical superiority of practitioners over preachers, the phenomenological movement (like the idealist) has been characterized by the converse – much preaching and little practice (Relph, 1981a, has pointed to the absence of substantive application of the phenomenological approach in geography):

> There is an essential difference between the contemplative intentions of this transcendental philosophy and the practical concerns of a social science, so that it is scarcely surprising that . . . geographers' . . . efforts have been directed towards the destruction of positivism as a *philosophy* rather than the construction of a phenomenologically sound *geography* (Gregory, 1978a, pp. 125–6).

There are some examples of the method's use, however. Tuan, in particular, has written several interpretative essays, and Relph, who 'would much prefer to see substantive applications rather than discussions of the possible uses of phenomenology' (1977, p. 178), has published his thesis on *Place and Placelessness* in which – implicitly, he says – phenomenological methods are used 'to elucidate the diversity and intensity of our experiences of place' (Relph, 1976, p. i): his essential themes are the sense of place and identity in the human make-up and the destruction of this through the growing placelessness of modern design. Other work generally quoted as phenomenological includes pieces on European settlement of the New World. Powell (1972), for example, has written on images of Australia and in his major work on the settlement of Victoria's western plains (Powell, 1970) has examined the conflict between official and popular environmental appraisals, the dialogue between these, and the learning process which resulted in the final settlement pattern (see also Powell, 1977, who presents the study of eiconics, or image-making, in the context of colonization processes). Billinge (1977), however, has queried whether all such works are really phenomenological:

> the idea has spread that since certain branches of our discipline are less susceptible to quantitative reduction (and, so the argument continues, by false extension, to *scientific* analysis), we can justify our partially formulated hypotheses, exploit the atypicality of our data, cease worrying about the validity of our reconstruction and within some weakly articulated framework label the whole exercise phenomenological (p. 64).

The study of perceived environments represents an 'important and vigorous movement' (p. 65) but, Billinge argues, phenomenology is not just the study of such environments: it is an approach to the study of those subjective sources which in its method is presuppositionless, concerned with the human consciousness and not only its output; thus 'phenomenological we have by no means become' (p. 67). In fact, much of the work which is claimed as phenomenological is probably closer to the idealist position, in that it does not attempt to investigate essences.

Closely associated with phenomenology is existentialism, and some geographers have experienced difficulty in separating the two (Entrikin, 1976). The basic difference is that phenomenology assumes the primacy of essence – the allocation of meanings results from the existence of consciousness – whereas to existentialists the basic dictum is 'being before essence – or 'man makes himself '. The process of defining oneself (creating an essence) involves creating an environment. Thus environments can be read as biographies, as the representations of man creating himself – 'for every landscape or every existential geography there is some one who can be held accountable' (Samuels, 1981, p. 131). Generalization may be possible in the analysis of such landscapes. Appleton (1975), for example, suggests that landscapes reflect two primal needs – prospect (the need to search for the means of survival), and refuge (the need to hide from threatening others) – although individuals and groups may seek to satisfy these needs in different ways. And Lowenthal (1975) has suggested that individuals seek to rewrite their biographies, and those of their ancestors, by their choices of what to preserve in the landscape.

All of this suggests that humanistic geography is concerned either with the study of individuals and their construction of phenomenal environments (as in Rowles, 1978) or with the analysis of landscapes as repositories of human meaning. As such, it may be considered separate from the subject matter of much human geography, notably behavioural geography and its investigation of everyday activity within environments (most of them man-made). But the phenomenological perspective has been adapted to such work, in the writings of Schutz on what he terms the 'taken-for-granted' world (Ley, 1977b). Much everyday behaviour is unconsidered, in that it involves no original encounters. The behaviour is habitual, because all of the stimuli encountered are processed as examples of particular types. Those types are not externally defined for the individual, but are created by him. The phenomenology of the taken-for-granted world is the study of those individually–defined typifications – of the unconsidered 'world of social reality' rather than 'a fictional non-existing world constructed by the scientific observer' (Ley, 1980, p. 10, quoting Schutz). It may be, of course, that inter-action within communities leads to common typifications. Quantitative methods may be used to identify the common elements, but as descriptive tools only. (Quantification is not tied to positivism, except when it is used to suggest laws and other generalizations.)

Humanistic geography is based on a profound critique of positivist work, both that which makes major assumptions about the nature of decision-making and that which seeks inductive laws of human behaviour that can be scientifically verified (Ley and Samuels, 1978; Powell, 1980). Its argument is for an understanding of man as a 'living, acting, thinking' being. To some, it is just a criticism – Entrikin (1976, p. 616) argues that:

humanist geography does not offer a viable alternative to, nor a presupposition-less basis for, scientific geography as it is claimed by some of its proponents. Rather the humanist approach is best understood as a form of criticism. As criticism the humanist approach helps to counter the overly objective and abstractive tendencies of some scientific geographers.

To others, however, the human condition can only be indicated by humanistic endeavour, for attitudes, impressions and subjective relations to places (the 'sense of place') cannot be revealed by positivist research.

To some critics, humanistic geography is concerned with relatively trivial matters and not with the major concerns of an applied geography concerned with improving the world. To Buttimer (1979, p. 30) such an applied geography is managerial, manipulating individuals and their environments rather than seeking to advance the process of 'human becoming'. To Relph (1981b, pp. 139–41) it implies geographism:

> the view that people should behave rationally in geographical, two-dimensional space . . . that cities and industries and transportation routes should be arranged in the most efficient way

which when used as a basis for planning

> will diminish the distinctiveness and individuality of . . . communities and places. Geographism involves the imposition of generalizations onto specific landscapes; it breeds uniformity and placelessness.

To him, as to Buttimer, planning should emphasize subjectivity and individuality. Scientists, engineers and planners may have sought to improve well-being, but in so doing have made man rootless and denied his individuality. Planning must be allied with environmental humility, by which

> places and communities would increasingly become the responsibility of those who live and work in them instead of being objects of professional disinterest (p. 201)

In this context, humanistic geography is not only a reaction against the dehumanizing treatment of man in spatial science and behavioural geography. It is also an argument against an applied geography which imposes that treatment on the landscape, and for a form of anarchism in which individuals are encouraged to realize what they are and how they can control themselves and their environments.

Conclusions

Two very different types of work have been reviewed in this chapter. The common thread linking them is that both are concerned with positive rather than normative investigations, with attempts to uncover how humans behave in the world rather than with contrasts between actual patterns of behaviour in space and those predicted from normative theories. Both types are part of a more general trend towards an anthropocentric focus within the social sciences, which in turn reflects reorientations in the external environment. From the mid-1960s on there was a growing disillusion with science and technology, especially among students, and the popularity of the social sciences boomed. Within the latter, there was a shift in emphasis from the aggregate to the individual, an increase in the relative volume of research conducted at the micro-scale, and growing unease about the role of social scientists in planning mechanisms. Both behavioural (or behaviourist) geography and the various humanistic approaches reflect these trends.

Beyond the general concern with the individual as a decision-maker there is little else to connect the two types of work discussed here. The former, as already pointed out, has maintained strong ties with the positivist/spatial-science tradition; data are collected from individuals, but these almost all concern the conscious elements in action; they are usually aggregated in order to allow statistically substantive and significant generalizations to be made about spatial behaviour, almost certainly in the context of the normative models of the spatial science school. In the humanistic strand, on the other hand, the intent has been to understand the individual and to re-recognize the dignity and humanity of man, but the output of substantive studies has not been great. This may indicate that the ethos of the positivist tradition has maintained a major constraint to the wholesale adoption of new philosophies and methodologies. Many investigations in human geography from 1970 onwards involved some behavioural concepts other than those of economic man, but relatively few human geographers have become humanistic geographers.

The two approaches are very much antithetical. One – behavioural geography – treats man as a responder to stimuli. It seeks to identify how different individuals respond to particular stimuli (and also how the same individual responds to the same stimulus in different situations), to isolate the correlates of those varying responses and to build models that can predict the probable impact of certain stimuli. The end-product is input to processes aimed at either providing environments to which people respond in a preferred way or at changing behaviour by changing the stimuli. (Such environmental planning is considered social control by some.) The other approach – humanistic geography – treats man as an individual constantly interacting with the environment and changing both himself and his milieu. It seeks to understand that interaction by studying it, as it is represented by the individual and not as an example of some scientifically–defined model of behaviour. And then by transmitting that understanding, it seeks to reveal man to himself and to enable him to develop the interactions in ways that will be self-fulfilling. The two cannot be combined, since the concept of behavioural laws is not in sympathy with the humanist view that men are alone responsible for their circumstances and destinies, and are free – if they so choose – to exercise that responsibility as they will. Few adherents of either approach totally derogate the other. Humanists recognize that behavioural geography and spatial science may be valid approaches to certain aggregate phenomena – such as trade flows (Ley, 1981; see also Tuan, 1977), and behaviouralists recognize that their methods are irrelevant to the study of emotions and aspects of landscape. (As suggested in the two main ways of studying space: Sack, 1980.) But the two have little in common, and are competing for central positions as the philosophy of *human* geography.

6

Relevance, liberals, and radicals

The period since the late 1960s has contained traumatic years in the countries being studied here. The underlying problem was economic uncertainty: after two decades of relatively high rates of economic growth and prosperity, the American and British economies began to experience serious difficulties. Further, it became increasingly clear that the prosperity of the previous decades had not been shared by all, and this was highlighted in the United States by the growing tempo of the Civil Rights movement. Student protest erupted in several countries during 1968, much though not all of it related to the Vietnam War, and increasing concern was being expressed about man's despoliation of his physical environment. Other issues included the relatively repressed state of women in western societies (this includes the geographical profession: Zelinsky, 1973a).

Much of the protest of the late 1960s was centred on particular issues and was relatively ephemeral. For many of those involved, the aim was to win reforms within society, in the classical liberal manner, while leaving its major structure untouched. For some, however, disillusion led to what Peet (1977) has termed a 'breaking-off' from liberalism and a move to more radical political stances. As he expressed it:

> The starting point was the liberal political social scientific paradigm, based on the belief that societal problems can be solved, or at least significantly ameliorated, within the context of a modified capitalism. A corollary of this belief is the advocacy of pragmatism — better to be involved in partial solutions than in futile efforts at revolution. Radicalization in the political arena involved, as its first step, rejecting the point of view that one more policy change, one more 'new face', would make any difference (p. 242).

Eventually, he says, the radicals exploded 'through the thick layers of ideology which in the most dangerous, mass-suicidal way, protect late capitalism' (p. 243) and settled on forms of socialism (see also Blaut, 1979). Their search for a perspective, and their debates with their more 'liberal' colleagues, form the material for this chapter.

Disenchantment and disillusion in academic geography

A forthcoming revolution in human geography, against the innate conservatism of behavioural studies, was foreseen by Kasperson (1971):

The shift in the objects of study in geography from supermarkets and highways to poverty and racism has already begun, and we can expect it to continue, for the goals of geography are changing. The new men see the objective of geography as the same as that for medicine – to postpone death and reduce suffering (p. 13).

This shift in objects and objectiveness is not easily defined, however. It is illustrated by the writings of one senior geographer, Wilbur Zelinsky (born in 1921), who was president of the Association of American Geographers (AAG) in 1973. His views may not be entirely typical, and are certainly more firmly stated than those of his peers, but they reflect a growing disillusion within geographical circles with past achievements (see also Cooke and Robson, 1976) and a wondering about future directions.

Zelinsky's (1970) first statement began:

> This is a tract . . . The reader is asked to consider what I have come to regard as the most timely and momentous item on the agenda of the human geographer: the study of the implications of a continuing growth in human numbers in the advanced countries, acceleration in their production and consumption of commodities, the misapplication of old and new technologies, and of the feasible responses to the resultant difficulties (p. 499).

He developed three basic arguments:
1   that man is inducing for himself a state of acute frustration and a crisis of survival;
2   that these conditions originate, and can only be solved, in the 'advanced' nations; and
3   that the current 'growth syndrome' has profound geographical implications.

Material accumulation can no longer be considered progress, he argues, for it is not sustainable; effort is currently misallocated on a massive scale, and there is a major geographical task involved in its sensible reallocation.

Zelinsky suggests five attitudes towards the problems of the growth syndrome:
1   ignore it;
2   accept that it will occur, eventually;
3   admit the existence of problems, but argue that are easily solved by the free market, perhaps with state guidance;
4   claim that nothing can be done, but that in any case we will survive; and
5   realize the potential for immediate, unprecedented trouble.

Zelinsky's own attitude is clearly the last of these, and he proposes three roles for the geographer in facing the oncoming disasters. The first – which involves a minimal political commitment and 'should not offend even the most rock-ribbed conservative scholar' (p. 518) – is the geographer as diagnostician, applying 'the geographic stethoscope to a stressful demography' (p. 519), mapping what he calls geodemographic load, environmental contamination, crowding and stress. The second involves the geographer as prophet, projecting and forecasting likely futures. Finally, there is the geographer as architect of utopia, educating with regard to problems and possible solutions and providing support for the unknown leaders who have the political will to guide society through the coming 'Great Transition'.

Ending his 'declaration of conscience' on a pessimistic note, Zelinsky argues regarding geographers,

> how woefully deficient we are in terms of practitioners, in terms of both quantity and quality, how we are still lacking in relevant techniques, but most of all that we are totally at sea in terms of ideology, theory and proper institutional arrangements (p. 529).

His criticisms are not confined to geographers, however; in his presidential address to the AAG (Zelinsky, 1975) he applied them to scientists en masse. Science, to Zelinsky, is the twentieth-century religion, but he claims it has failed to avert the oncoming crisis. Its disciplinary specialisms and separatism 'fog perception of larger social realities' (p. 128), while 'fresher, keener insights, along with much better prose' (p. 129) are produced by the brighter contemporary journalists.

Zelinsky identifies five crucial axioms as foundations of science:
1 that the principle of causality is valid for studying all phenomena;
2 that all questions are soluble;
3 that there is a final state of perfect knowledge;
4 that findings have universal validity; and
5 that total scientific objectivity is possible.

But the social sciences have failed to live up to these, he claims, for several reasons: their immaturity; their use as a refuge for mediocre personnel; the difficulty of their subject matter concerning interpersonal relationships; their problems of observation and experimentation; and the political and other problems involved in applying their proposed solutions. But furthermore, their major cause for failure is the irrelevance of the natural science model to the study of society:

> If we are in pursuit of nothing more than information or knowledge, then there is some value in copying the standard formula of a research paper in the so-called hard sciences. . . . But if we are in pursuit of something more difficult and precious than just knowledge, namely understanding, then this simple didactic pattern has very limited value (p. 141).

Zelinsky's views were echoed in another presidential address to the same body, with Ginsburg (1973) writing that:

> Much so-called theory in geography . . . is so abstracted from reality that we hardly recognize reality when we see it. . . . The increasing demand for rigour to cast light on trivia has come to plague all of the social sciences . . . the most important questions tend not to be asked because they are the most difficult to answer (p. 2).

At about the same time, the Association established a Standing Committee on Society and Public Policy (Ginsburg, 1972), which White (1973) hoped would 'be alert to distinguishing the fatuous problems and the activities that are pedestrian fire-fighting or flabby reform' (p. 103). This statement was made at a session on geography and public policy at the Association's 1970 conference; a similar theme was chosen for the 1974 conference of the Institute of British Geographers (IBG). It should be noted, however, that not all geographers accepted White's aim – using the AAG to influence public policy – whilst not denying the value of geographic method in social engineering. According to Trewartha (1973):

> I must demur when he proposes that it should be a corporate responsibility of our professional society to become an instrument for social change. . . . From the beginning, the unique purpose of the Association of American Geographers has been to advance the cause of geography and geographers; it was never intended to be a social-action organization. . . . All kinds of research, pure as well as applied, should be equally approved and supported by the AAG (p. 79).

Two major arguments can be identified in the materials just reviewed: that geographical research should be relevant to major societal problems; and that the positivist-based spatial-science methodology may be inappropriate for such a task. As several have been quick to point out, neither of these concerns was particularly new, especially the first: House (1973), for example, has reviewed the tradition of involvement in public policy by British geographers, and Stoddart (1975b) has identified the late nineteenth-century views of Réclus and Kropotkin as 'the origins of a socially relevant geography' (p. 190) – the latter was later rediscovered by the 'radicals' (Peet, 1978). Nor was the more 'revolutionary' approach of those who had 'broken off' from liberalism particularly novel. Santos (1974), for example, has reminded English-speaking geographers of the marxist-inspired works of Jean Dresch on capital flows in Africa and of Jean Tricart on class conflict and human ecology, both prior to the Second World War; an English geographer, Keith Buchanan, was well known for his left-wing opinions in New Zealand during the 1960s, and his presidential address on the need for studying 'the absolute geographical primacy of the state; especially in the non-Western world' (Buchanan, 1962), produced an acid response from Spate (1963).

Relevance, on what, and for whom?

The claims that geographical work should be more relevant to major societal problems raised queries about the nature of that relevance, and it soon became apparent that there was no consensus on what should be done, and why. The ensuing debate is illustrated by a number of contributions to the British journal *Area* during the early 1970s.

The opening statement was by Chisholm (1971b), who identified differences between governments, with their interests in cost-effective research and their primacy in decision-making, and academics, some of whom are concerned to protect their academic freedom and their right to be the sole judges of what they study and publish. Traditionally, geographers had advised governments as information gatherers and 'masterful synthesizers' and had not been involved in the final stages of policy-making: they had been delvers and dovetailers, but not deciders. On the latter role, unfortunately:

> The magic of quantification is apt to seem rather less exciting when the specifications of the goods it can deliver are inspected at close quarters (p. 66). . . . The danger with empirical science is the absence of guidance at the normative level as to which of various options one should take (pp. 67–8).

The challenge to human geography, according to Chisholm, was to define such norms.

To Eyles (1971), the focus of relevant research should be 'some of the social and spatial inequities in society' (p. 242), and the challenge is to study the distribution of power in society, which is the mechanism that allocates scarce resources. Research would then identify the disadvantages of relative powerlessness – and would provide the basis for policy which redistributes resources.

Also in 1971, British readers were introduced to the ongoing debates in American geography with two reports on the 1971 AAG conference. This was not the first at which major social issues had been raised: the 1969 meeting should have been in Chicago, but it was transferred to Ann Arbor as a protest over events at the 1968 Democrat Party Convention in Chicago. At Ann Arbor, the radical journal *Antipode* was launched, and those attending were confronted by the problems of the inhabitants of Detroit's black ghetto. By the 1971 meeting, 'many geographers were deeply frustrated by a sense of failure' (Prince, 1971b, p. 152) to deal with major social issues, but while some members were taking notice of 'the sufferings of the outside world' (p. 152) other scholars remained 'locked in private debates, preoccupied with trivia, mending and qualifying accepted ideas' (p. 153).

In his report of the 1971 conference, Smith (1971) suggested that American geography was about to undergo another revolution, to counter a situation in which 'geography is overpreoccupied with the study of the production of goods and the exploitation of natural resources, while ignoring important conditions of human welfare and social justice' (p. 154). This forthcoming revolution would involve fundamental revaluation of research, teaching activities, and basic social philosophies, and was represented at the conference by activities such as the foundation of SERGE (The Socially and Ecologically Responsible Geographer) by Zelinsky and others and by a motion at the Annual General Meeting condemning United States involvement in Vietnam. Smith was unsure, however, whether this revolution, with its emphasis on social as against economic concerns, would spread to Britain:

> The conditions which have helped to spawn radical geography in the United States include the existence of large oppressed racial minorities, inequalities between rich and poor with respect to social justice, a power structure and value

system largely unresponsive to the needs of the underprivileged, and an unpopular war which is sapping national economic and moral strength. These conditions do not exist in Britain or exist in a less severe form, and the stimulus for social activism in geography is thus considerably less than in America (pp. 156–7).

Dickenson and Clarke (1972) responded that British geographers had long been concerned with 'relevant' issues, with particular respect to the Third World.

Another commentary on the 1971 conference of the AAG at Boston came from Berry (1972c), who felt that he had observed just a new fad involving 'new entrants to the field seeking their "turf" '. He could identify no real commitment:

> The majority of the new revolutionaries, it seems, are essentially 'white liberals', quick to lament the supposed ills of society and to wear their bleeding hearts like emblems or old school ties – and quicker to avoid the hard work that diagnosis and action demand. A smaller group of hard-line marxists keeps bubbling the potage of liberal laments. *In neither group is there any profound commitment to producing constructive change by democratic means. . . . If either of these will be the 'new geography' of the 1970s, count me out* (pp. 77–8).

To him, the academic geographer should provide a base of knowledge on which policy can be built, and this involves a close involvement with the education of future generations of policy-makers. But to Blowers (1972), 'The issue is not how we can cooperate with policy-makers, but whether and in what sense we should do so. It is a question of values' (p. 291). To him, the activities proposed by Berry would be strongly supportive of the status quo, and unlikely to produce fundamental social reform. Smith (1973a) responded to Berry that 'bleeding hearts sometimes help to draw attention to important issues, and marxists can make valuable contributions in the search for alternatives to existing institutions and policies' (p. 1) and pointed out that the current 'fad' was no more pronounced than that of the quantifiers a decade earlier. The results of that earlier 'revolution' offered little for the solution of social problems, however, and Smith doubted the value of the large projects established by the AAG as part of its geography and public policy drive. Instead the research focus should highlight particular problems, while teaching should have emphases on 'a man in harmony with nature rather than master of it, on social health rather than economic health, on equity rather than efficiency, and on the quality of life rather than the quantity of goods' (p. 3). Chisholm (1973) advocated caution in the corridors of power, because geographers had done insufficient substantive research to back up a 'hard sell'; Eyles (1973, p. 155) argued that any entry to those corridors 'assumes that the structure underlying policy alternatives is basically sound'; and Blowers (1974) wrote that in the corridors one can only influence, not decide, and that for the latter task geographers must develop their political convictions and act accordingly.

This debate on if, and how, geographers should contribute to the solution of societal problems was a major item at the 1974 annual conference of the IBG. In his presidential address, Coppock (1974) presented the challenges, opportunities and implications of geographical involvement in public policy, an involvement which he felt the current generation of students welcomed. Policy-makers, he felt, were largely ignorant of the potential geographical contribution, while at the same time geographers seemed unaware that 'there is virtually no aspect of contemporary geography which is not affected to some degree by public policy' (p. 5). Coppock sought to change this, to have geographers identify the contributions that they could make, to encourage research relevant to those contributions, and to enter a dialogue with those who advise on and implement public policy.

Other conference contributors were neither as optimistic nor as committed as Coppock. Hare (1974), himself an adviser to the Canadian government, reacted to cries that geographers were not consulted enough with the reply 'Thank goodness'. His argument was that geography, as a discipline, is irrelevant to the separate domain of public-policy-making, although geographers, as individuals, because of the breadth of their training, could offer much that was valuable: his conclusion was 'Geography no, geographers yes' (p. 26). His was a different response to mounting social concerns from that of Steel (1974) who told the British Geographical Association:

> At geographers we often get hot under the collar over the number of theoretical economists who are called on to advise the governments of developing countries. We comment on how much better World Bank surveys of countries would be if they were prepared, at least in part, by geographers. . . . We wonder why university departments of geography are not engaged on a consultancy basis more often than they are, and we marvel that the Overseas Development Administration in London has only a handful of geographers on its staff where, we feel, an army would be more appropriate (p. 200).

In a later paper, Hare (1977) argues that a major reason for a lack of geographical contributions to public policy in recent years may be the poverty of their training: in recent years we have 'swept geography departments into the social-science divisions of faculties of arts and sciences where, from playing second fiddle to geologists or literary critics, we learned to play second fiddle to economists and sociologists' (p. 263). Geographers, he argued, would have to rebuild their discipline based on the centrality of man-environment interactions, with a new brand of physical geography that leans heavily on biological ideas and sources. Thus:

> We must reassert the old, essential truth that geography is the study of the earth as the habitat of man, and not some small sub-set of that gigantic theme (p. 266)

and regional syntheses should be stressed, because

> Realizing that regional perspectives are necessary in politics is to get one's manhood back (p. 269).

(See also Steel, 1982; Hart, 1982.)

Hare's views on the current irrelevance of geography to public policy were supported by Hall (1974) who argued that

> geography, most clearly of all the social sciences, has neither an explicit nor an implicit normative base . . . spatial efficiency . . . is rather a description of what men seek to do in actuality . . . not . . . any objective to be achieved or objective function to be maximized. (p. 49).

Policy-makers must seek their norms elsewhere; geographers, meanwhile, must develop a new political geography which will aid in their understanding of the crucial role of political decisions in structuring spatial systems (Johnston, 1978b).

Two other papers given at the 1974 IBG conference argued against Coppock's programme. Leach (1974), for example, claimed that governments, as paymasters, already constrained what geographers could do research on, and as a result geographers were being used; their only alternative was political action. Harvey's (1974c) contribution was entitled 'What kind of geography for what kind of public policy?'. Individuals wishing to become involved in policy-making were, he argued, stimulated by motives such as personal ambition, disciplinary imperialism, social necessity and moral obligation: at the level of the whole discipline, on the other hand, geography had been coopted, through the universities, by the growing corporate state, and geographers had been given some illusion of power within a decision-making process designed to maintain the status quo. Indeed, to Harvey the corporate state is 'proto-fascist' (p. 23), a transitional step on the path to the barbarism of Orwell's *1984*. The function of academics, he claimed, was to counter such trends, to expunge the racism, ethnocentrism and condescending paternalism from within their own discipline and to build a humanistic subject and thereby assist all human beings 'to control and enhance the conditions of our own existence' (p. 24).

Harvey (1973) had argued previously that the current mode of analysis in geography offered little for the solution of pressing societal concerns:

> There is an ecological problem, an urban problem, an international trade problem, and yet we seem incapable of saying anything of depth or profundity about any of them. When we do say anything, it appears trite and rather ludicrous. . . . It is the emerging objective social conditions and our patent inability to cope with them which essentially explains the necessity for a revolution in geographic thought (p. 129).

He recognized three types of theory:

1 status quo, which represents reality accurately but only in terms of static patterns, and therefore cannot make predictions which will lead to fundamental social change;

2 counter-revolutionary, which also represents reality, but obfuscates the real issues − it is 'a perfect device for non-decision making, for it diverts

attention from fundamental issues to superficial or non-existent issues' (p. 151); and

3 revolutionary, which is grounded in the reality it seeks to represent, and is formulated so as to encompass the contradictions and conflicts which produce social change.

Harvey clearly wished to write revolutionary theory, thereby overthrowing the current paradigm; his blueprint for geography

> does not entail yet another empirical investigation. . . . In fact, mapping even more evidence of man's patent inhumanity to man is counter-revolutionary in the sense that it allows the bleeding-heart liberal in us to pretend we are contributing to a solution when in fact we are not. This kind of empiricism is irrelevant. There is already enough information. . . . Our task does not lie here. Nor does it lie in what can only be termed 'moral masturbation' of the sort which accompanies the masochistic assemblage of some huge dossier of the daily injustices. . . . This, too, is counter-revolutionary for it merely serves to expiate guilt without our ever being forced to face the fundamental issues, let alone do anything about them. Nor is it a solution to indulge in that emotional tourism which attracts us to live and work with the poor 'for a while'. . . . These . . . paths . . . merely serve to divert us from the essential task at hand.
>
> This immediate task is nothing more nor less than the self-conscious and aware construction of a new paradigm for social geographic thought through a deep and profound critique of our existing analytical constructs. This is what we are best equipped to do. We are academics, after all, working with the tools of the academic trade . . . our task is to mobilize our power of thought, which we can apply to the task of bringing about a humanizing social change (pp. 144–5).

To Harvey, then, relevant geography involves the revision of geographic theory; such new theory will be built on a marxist base, and its dissemination will achieve social reform through the education process. (This gradualist view to reform through education is only implicit in Harvey's work; Johnston, 1974, p. 189.) To Blaut (1979), the benefit of marxist theory was that it can handle two crucial issues that positivist theory cannot: increasing injustice and heightened economic and social instability.

There is a long tradition of liberal contributions by geographers, both in their research and associated activities and in their teaching (see House, 1973, and Hall, 1981b, for reviews). In Britain, applied geography – as information gathering and synthesizing (p. 163) – has a long and substantial record. Stamp's Land Utilization Survey of Britain in the 1930s, his involvement in the preparation for post-war land-use planning, and the use of geographical material in that preparation, provide a major example (Stamp, 1946). Other aspects of land-use planning were of interest to geographers in the 1930s and 1940s, and were discussed several times at the Royal Geographical Society (Freeman, 1980b); there was similar geographical involvement in the United States (Kollmorgen, 1979). After the Second World War – during which geographers made many contributions as information synthesizers and gatherers, the latter including the development of air-photo interpretation – land-use planning was established on a large scale, and trained geographers provided a large proportion of its personnel. Many academic geographers remained active in applied work. Technical developments in cartography and data handling have seen them involved in redistricting for Congressional elections (Morrill, 1981), a wide range of mapping activities (Rhind and Adams, 1980), and developing regional bases for the presentation of Census statistics (Coombes *et al.*, 1982), for example. Policies have been evaluated (e.g. Keeble, 1976), and much of the entropy-maximizing systems modelling (p. 119) was developed to provide procedures for the joint activities of land-use and transport planning (Wilson, 1974).

Most of this work was set in an (at least implicit) positivist framework. The planning of agricultural land use, for example, assumed a relationship between the physical environment and agricultural productivity, and that on the location of industry was based on certain beliefs regarding spatial efficiency. As the positivist approach became more explicit, so the claims to be able to make a scientific contribution to land-use and spatial organization planning increased. This occurred in the 1960s; most of it was pragmatic, but there has been some attempt to develop a theory of decision-making relevant to applied geography (Hall, 1981a, 1982). The change in the 1970s, at least to some, was in the substantive focus, not the method. But there was a polarization of views developing. On the one hand, there were those who advocated greater commitment to applied geography (or to 'geography and public policy'), because of both the pressing nature of societal problems and the need for geography to be 'relevant' in such hard times. On the other hand, there were those who advocated more radical reappraisals of the discipline's role, based largely on socialist rather than capitalist ideologies. (There were also those who were unconvinced of the need to leave the academic ivory tower.) The major contributions from these two groups are reviewed in the next sections.

The liberal contribution

In the present context, liberalism is defined as the idea 'shared by many modern American liberals who, characteristically, combine a belief in democratic capitalism with a strong commitment to executive and legislative action in order to alleviate social ills' (Bullock, 1977, p. 347). Liberals, then, are concerned that all members of society do not fall below certain minimum levels of well-being (variously defined), and are prepared for state action within the capitalist structure in order that this can be achieved. Within geography, much of work conducted in this ethos has focused on description rather than on theory-construction: the few exceptions include Chisholm's (1971a) investigation of the potential of welfare economics as a basis for normative theory which does not involve profit-maximization goals (see also Wilson, 1976b).

Mapping welfare

During the 1960s, a lot of research was reported, under the general title of factorial ecologies (p. 84), on the application of multivariate statistical procedures to large data matrices representing spatial variations in population characteristics. These works, according to Smith (1973b, p. 43), were over-reliant on certain types of census data and therefore provided little information on social conditions. There had been some earlier attempts to structure analyses of such data towards particular ends, as in the work of rural sociologists on farmers' levels of living (Hagood, 1943), a concept introduced to the geographical literature by Lewis (1968), and Thompson *et al.*'s (1962) investigation of variations in levels of economic health among different parts of New York state. Only in the 1970s, however, was the factorial ecology set of procedures adapted to the task of mapping social welfare on any significant scale.

Two workers led in this adaptation. For Knox (1975), a fundamental objective for geography is to map social and spatial variations in the quality of life, both as an input to planning procedures and as a means of monitoring policies aimed at improving welfare. The concept of level-of-living was divided into three sets of variables for this reporting task – physical needs (nutrition, shelter and health); cultural needs (education, leisure and recreation and security); and higher needs (to be purchased with surplus income) – and statistical procedures were used to provide accurate portrayals of the spatial variations in meeting these needs. With the resultant maps, the geographer must then decide whether he is playing a sufficient role in awakening human awareness of the extent of the disparities or 'whether he is under an obligation to help society improve the situation' (p. 53).

Smith's (1973b) very similar work was set in the context of the American social indicators movement and the growing belief there that GNP and national income 'are not necessarily direct measures of the quality of life in its broadest sense' (p. 1). His aim was to initiate the collection and dissemination of territorial social indicators, to point out the extent of discrimination by place of residence which occurs in the United States. Again, multivariate statistical procedures were employed to provide the needed maps, at inter-state, inter-city and intra-urban scales. (See also Cox, 1979, who extended the treatment to the international scale.) With these, it was necessary for geographers to get involved in the political struggle, to be social activists as well as social reporters.

To use Chisholm's terms (p. 163), these two works represented the geographer as delver and dovetailer, as a provider of information on which more equitable social planning could be based. Several other studies of individual elements in that planning performed similar roles, and also suggested spatial policies which could lead to social improvements. Harries (1974), for example, studied spatial variations in crime rates and the administration of justice, and argued that predictive models of criminal patterns, in the positivist mould, could aid in the organization of police activities; Shannon and

Dever (1974) investigated variations in the provision of health-care facilities and argued for spatial planning which would improve the services offered to the sick (which is different from a geography of prophylaxis: Fuller, 1971); and Morrill and Wohlenberg (1971) studied the geography of poverty in the United States, proposing both social policies – higher minimum wages, guaranteed incomes, guaranteed jobs, and stronger anti-discrimination laws – and spatial policies – such as an extensive programme of economic decentralization to a network of regional growth centres – which would alleviate this major social problem.

An alternative, and highly personal, programme of mapping variations in human welfare was advanced by Bunge, who had prepared a 'geobiography' of his home area, part of the black ghetto of Detroit (Bunge, 1971). His is a deeply humanitarian concern for the future of mankind, which he interprets as a need to ensure a healthy existence for children; he wants a 'dictatorship of the children' (Bunge, 1973b, p. 329) with regions – 'may the world be full of happy regions' (p. 331) – designed for them. This requires a reduction in the worship of machines, which are inimical to children's health (Bunge, 1973c) and a mapping of the sorts of variables never collected by external agencies and therefore requiring the development of geographical expeditions within the world's large cities; these maps would include roach regions, parkless spaces, toyless regions, and rat-bitten-children regions (Bunge, 1973d), and some have been prepared for Detroit and for Toronto (Bunge and Bordessa, 1975).

Attempts at understanding

The mapping investigations just discussed are very largely descriptive, and any prescriptions offered are based on limited theoretical foundations. Other investigations included attempts to develop the necessary theoretical understanding. Cox (1973), for example, looked at the urban crisis in the United States – the racial tensions and riots, municipal bankruptcies, and the role of the government in the urban economy – presenting his analysis in terms of conflict over access to sources of power. This was intended as part of an educational exercise, for:

> It would be utopian to think that we can propose solutions on the basis of our analysis. The locational problems and locational consequences of policies weave too intricate a web for that to be possible. All we can hope to do is inform. To be aware of the problems and of their complexity may induce some sensitivity in a citizenry which has shown as yet precious little tolerance for the other point of view (p. xii).

Nevertheless, his final chapter is entitled 'Policy implications' in which are discussed two imperatives towards greater equity in the provision of public services – the moral imperative and the efficiency imperative (the latter applies to the total level of welfare in society as well). The policies presented involve spatial reorganization to achieve the desired equity, including metropolitan integration, community control, population redistribution, and transport improvements. A similar focus on spatial reorganization is provided in Massam's (1975) review of geographical contributions to social administration: his evaluation of service provision is in terms of the spatial variables of distance and accessibility, with major chapters on the size and shape of administrative districts and on the efficient allocation of facilities within such districts (see also Hodgart, 1978).

Cox's work heralded an increased geographical interest in a much neglected field, the role of the state in capitalist society (see also Cox, Reynolds and Rokkan, 1974; Dear and Clark, 1978; Johnston, 1982a; Cox, 1979). Traditionally, political geography had been concerned with the state at the macro-scale only, dealing with political regions and boundaries and with the operations of the international political system (e.g. Muir, 1975). The similarly underdeveloped field of electoral geography had highlighted spatial variations in voting, but there had been little work on both the geographical inputs to voting and the geographical consequences of the translation of votes into political power (Taylor, 1978; Taylor and Johnston, 1979). The state is involved in many aspects of economic and social geography, however, as both Buchanan (1962) and Coppock (1974) have stressed, but few geographers have investigated this involvement in any detail, or the electoral base on which it is founded (Brunn, 1974; Johnston, 1978b).

An attempt to present an understanding of spatial variations in well-being was presented by Coates, Johnston and Knox (1977). After defining the components of well-being and mapping their variations at three scales – international, intra-national, and intra-urban – they introduced three sets of causes of such variations: the division of labour; accessibility to goods and facilities; and the political manipulation of territories. (See also Cox, 1979.) Finally, they evaluated spatial policies aimed at the reduction of spatial inequalities, such as various forms of positive discrimination by areas. Their conclusions were that, of their three sets of causes, the division of labour is the primary determinant of levels of social well-being. Creation of this division is a social and not a spatial process, though it has clear spatial consequences, so that:

> The root causes of spatial inequalities cannot be tackled by spatial policies alone, therefore. Inequalities are products of social and economic structures, of which capitalism in its many guises is the predominant example. Certainly inequalities can be alleviated by spatial policies . . . but alleviation is not cure: whilst capitalism reigns, however, remedial social action may be the best that is possible . . . the solution of inequalities must be sought in the restructuring of societies (pp. 256–7).

Hägerstrand (1977) was drawn to similar, if not more pessimistic, conclusions:

> When the world is stable and/or unhampered liberalism prevails, then there is probably not much to do for geographers except surviving in academic depart-

ments trying to keep up competence and train schoolteachers in how wisely arranged the world is (p. 329).

(Hägerstrand's definition of liberalism in this case is undoubtedly that of economic liberalism, which favours price competition in the 'marketplace' and frowns on 'state interference': Brittan, 1977, p. 188).

A more ambitious attempt at explaining spatial variations in well-being is Smith's (1977) book, in which he argues that:

> the well-being of society as a spatially variable condition should be *the* focal point of geographical enquiry . . . if human beings are the object of our curiosity in human geography, then the quality of their lives is of paramount interest (pp. 362–3)

and he acts on this by essaying

> a restructuring of human geography around the theme of *welfare* . . . to provide both positive knowledge and guidance in the normative realm of evaluation and policy formulation (p. ix).

The book proceeds from theory through measurement to application. The theoretical section is an amalgam of normative welfare economics with marxian perspectives on the creation of value plus the political conflict for power.

> The analysis will inevitably reveal certain fundamental weaknesses of the contemporary capitalist-competitive-materialistic society, but the temptation to offer a more radical critique of existing structures has been resisted, in favour of an approach that builds on the discipline's established intellectual tradition (p. xi).

Two types of solution to perceived spatial inequalities are identified: liberal intervention, and radical, structural reform – the former is emphasized because 'most social change is incremental rather than revolutionary' (p. xii). Thus one concluding chapter – entitled 'Spatial Reorganization and Social Reform' – emphasizes reorganization of administrative areas, although geographers are later urged to be 'open-minded enough to see that there may be greater merit to some of the development strategies practised with evident if not painless success in the USSR, China, Cuba and so on' (p. 358). Further, counter-revolutionary possibilities are foreseen – 'overpreoccupation with spatial reorganization may serve the interests of the existing rich and powerful by helping to obscure the more fundamental issues' (p. 359) – and some socialist arguments are forthcoming – 'That spaces and natural resources should be privately owned, with their use subject to the chances of individual avarice, altruism or whim, is increasingly an anachronism' (p. 360). The conclusion is that:

> As geographers we have a special role – a truly creative and revolutionary one – that of helping to reveal the *spatial* malfunctionings and injustices, and contributing to the design of a spatial form of society in which people can be

really free to fulfil themselves. This, surely, would be progress in geography (p. 373).

Environmentalism

The late 1960s was a period of rapidly increasing concern about environmental problems; as Mikesell (1974) described it for the United States:

> Towards the end of the 1960s the American public was overwhelmed with declarations of an impending environmental crisis. . . . Since that time, crisis rhetoric and a yearning for simple answers to complicated questions has given way to a more sophisticated and deliberate search for environmental understanding. Ecology has been institutionalized (p. 1).

Two of the leaders of the public debate disagreed as to the cause of the problems (O'Riordan, 1976, pp. 65–80): Ehrlich argued for the primacy of population growth, and popularized the concept of zero population growth; Commoner claimed that technological advances and the consequent rapid depletion of resources plus deposition of pollutants created the major problems.

Both of these arguments have clear geographical components, as Zelinsky and others realized (p. 160), and geographers have a considerable record of activity in resource conservation. In the United States, for example, George Perkins Marsh has written on the topic in 1864 (Lowenthal, 1965) and in the 1930s the climatologist Warren Thornthwaite had been closely involved in the soil conservation movement established as a consequence of the Dust Bowl phenomenon. Their interest in landscape modification was advanced by Sauer and his followers, and reflected in the symposium *Man's Role in Changing the Face of the Earth*: there was similar interest elsewhere, as exemplified by Cumberland's (1947) pioneering classic on soil erosion in New Zealand. Nevertheless, Mikesell wrote that 'developments in geography have been such that the several phases of national preoccupation with environmental problems have not produced a general awareness of our interests and skills' (p. 2).

As part of the AAG's increased commitment to public affairs, its Commission on College Geography established a Panel on Environmental Education and sponsored a Task Force on Environmental Quality. The latter reported (Lowenthal *et al.*, 1973) that geographers would make excellent leaders for the educational tasks in hand because:

1 of the breadth of their training and their ability to handle and synthesize material from a range of sources;
2 their acceptance of the complexity of causation;
3 the range of information which they are trained to tap;
4 their interest in distributions; and
5 their long tradition of study in this area.

All of these had fostered the development of expertise in work on environmental perception, on vegetation succession, and on relationships between land use and soil erosion, which could be used as the bases for environmental impact statements, the elaboration of environmental choices, and international research collaboration.

Geographical efforts in the area of man-environment interrelationships were of two types. The first, traditionally geographical, was concerned with description and analysis. Review volumes such as *Perspectives on Environment* (Manners and Mikesell, 1974) were prepared, and a particular interest in problems of the physical environment of urban areas was generated (Detwyler and Marcus, 1972; Berry and Horton, 1974; Berry *et al.*, 1974). The second type of effort focused more precisely on issues of environmental management (O'Riordan, 1971a, 1971b), with particular emphasis on its economic aspects: as Kates (1972, p. 519) pointed out, economics provided the theories and prescriptions of the 1960s. A topic of especial interest was the study of leisure, of the growing demand for recreational facilities, and the impact of recreational activities on the environment (Patmore, 1970).

Despite such activity, Mikesell concluded in 1974 that the geographical contribution to environmentalism had not been great. Regarding the prognostications of *The Limits of Growth* (Meadows *et al.*, 1972), for example, he commented that 'the debate on this most relevant of all issues has attracted remarkably little attention from geographers' (Mikesell, 1974, p. 19) – though see Eyre (1978) – and he concluded more generally that: 'one must add hastily that many of the environmental problems exposed in recent years and also many of the social and philosophical issues debated during the environmental crusade have not been given adequate attention by geographers' (p. 20). Such a conclusion is supported by perusal of the contents of recent geographical journals, and of O'Riordan's (1976) lengthy bibliography.

The breadth of study in modern environmentalism is indicated by O'Riordan's volume. Much of the work is set in the liberal humanitarian tradition already illustrated in this section. O'Riordan draws four conclusions: environmentalism challenges many aspects of Western capitalism; it points out paradoxes rather than clear solutions; it involves a conviction that better modes of existence are possible; and it is a politicizing and reformist movement, based on a realization of the need for action in the face of impending scarcity and a lack of faith in the western democracies (pp. 300–1). A new social, environmental order is required. O'Riordan identifies three possibles – centralized, authoritarian, and anarchist; his choice is the liberal one, for the middle-of-the-road:

> we must individually and collectively seize the opportunities of the present situation to end the era of exploitation and enter a new age of humanitarian concern and cooperative endeavour with a driving desire to re-establish the old values of comfortable frugality and cheerful sharing (p. 310).

This new era, which will involve a new political order based on a combination of local self-determination and supranationalism, can be achieved through education – environmental education will form a preparation for citizenship.

To some writers, the revival of interest in environmental issues – mainly through the study of resources and their management – provides a contemporary linking of human and physical geography. Relatively little of the research and textbook writing indicates any integration of the two, however (Johnston, 1983a), because the focus is almost invariably on the processes studied in one of the sub-disciplines only. Thus physical geographers study man as a catalyst for environmental processes, as a result of trends such as demographic growth, technological sophistication, urbanization, and demands for resources, but they take those trends for granted and do not ask what processes generate them. Similarly, human geographers take the resources of the physical environment as given and do not inquire of their genesis. Thus while bridges have been built across the human-physical geography interface, there has been no integration of the study of physical and societal processes; for human geographers their links with other social scientists are very much stronger than those with environmental scientists (as illustrated, for example, by the lack of references to the physical geography literature in O'Riordan, 1976).

Radical alternatives

Peet (1977) has pointed out that the early 'radical' work by geographers in the late 1960s was liberal in its attitudes (p. 142):

> Radicals investigated only the surface aspects of these questions – that is, how social problems were manifested in space. For this, either we found the conventional methodology adequate enough or we proposed only that 'existing methods of research must be modified to some extent if they are to serve the analytic and reconstructive policies of . . . radical applications' (Wisner, 1970, p. 1) . . . we were fitting into an established market . . . we were amenable to established ways of thinking . . . we were useful in providing background ideas for the formulation of 'pragmatic' public policy directions, and so could not, and were not, engaging in radical analysis and practice (p. 245).

This was illustrated by his own paper on poverty in the United States (Peet, 1971) – in which, like Morrill and Wohlenberg (see p. 170), he argued for a series of growth centres in the poverty areas – and also by the tenor of the articles in early issues of the radicals' journal, *Antipode*. Morrill (1969), for example, argued

> against the 'New Left' premise that a revolution is the only route to progress . . . the dreams of revolution are naive . . . the New Left vastly exaggerates potential support . . . a 'revolutionary program' is hopelessly dated and simplistic . . . the 'New Left' underestimates the capacity of our society for change. . . . All revolutions seem to have been betrayed by incompetents who preferred exercising power to executing reform (pp. 7–8)

and later (Morrill, 1970b) that

> A simple marxist-type change in the ownership of business from private to a
> government (or union) bureaucracy would in all probability decrease produc-
> tion, and would not necessarily bring any improvement in basic conditions. The
> key is to retain the institution of private property while instituting social control
> over its exchange and circumscribing its power over people (p. 8).

The case for a marxist approach was first presented formally by Folke
(1972) in a critique of Harvey's (1972) paper on ghetto formation and
counter-revolutionary theory. To Folke, geography and the other social
sciences are 'highly sophisticated, technique-orientated, but largely descriptive disciplines with little relevance for the solution of acute and seemingly
chronic societal problems . . . theory has reflected the values and interests of
the ruling class' (p. 13). Liberal arguments such as Morrill's are dismissed as
unlikely to succeed. Morrill had argued in his two papers in *Antipode* for
change to be brought about by persuasion, producing a capitalist-socialist
convergence. But this is the social democrat method practised in Sweden

> where it has been shown over and over again that the idea of equal influence for
> employers and employees is an illusion. After half a century of social-democratic
> rule injustices and inequalities still prevail. . . . No small group of experts can
> accomplish anything . . . when it runs counter to the interests of the dominant
> social forces. These are not interested in equality or justice, but in profit (Folke,
> 1972, p. 15).

Radical change requires mass mobilization, and so, to Folke, Harvey's call
for a new paradigm within geography was insufficient. What is needed is a
new paradigm for a unified social science, containing geography, which
deals with problems in all their complexity and provides not only theory but
the basis for action: 'Revolutionary theory without revolutionary practice is
not only useless, it is inconceivable . . . practice is the ultimate criterion of
truth' (p. 7). (See also Eliot Hurst, 1980.)

The major contribution to the case for a marxist-inspired, materialist
theory-development within geography was made by David Harvey, notably
in his book of essays *Social Justice and the City* (Harvey, 1973). The book is
presented as autobiographical, illustrating the evolution of Harvey's views
towards an acceptance of Marx's analysis:

> as a guide to enquiry . . . I do not turn to it out of some *a priori* sense of its
> inherent superiority (although I find myself naturally in tune with its general
> presupposition of and commitment to change), but because I can find no other
> way of accomplishing what I set out do or of understanding what has to be
> understood (p. 17).

The first part of the book is entitled 'Liberal Formulations' and comprises
essays which analyse problems of inequality within societies in terms of the
mechanisms which allocate income; the role of accessibility and location in
those mechanisms is stressed. This leads him to an attempt to define territo-

rial social justice, which separates the processes allocating incomes from those which produce them. Only in the second part of the book – 'Socialist Formulations' – is it

> finally recognized that the definition of income (which is what distributive justice is concerned with) is itself defined by production. . . . The collapse of the distinction between production and distribution, between efficiency and social justice, is a part of that general collapse of all dualisms of this sort accomplished through accepting Marx's approach and technique of analysis (p. 15).

The transition in Harvey's approach is marked by the paper on the ghetto (Harvey, 1972). He begins with a critique of Kuhn's model of scientific development (p. 19), asking how anomalies to the current paradigm arise and how they are translated into crises. The problem with Kuhn's analysis, he claims, is that it assumes that science is independent of its enveloping material conditions, when in fact it is very much geared to its containing and constraining society. Recognition of this point is important for geographers because:

> the driving force behind paradigm formulation in the social sciences is the desire to manipulate and control human activity in the interest of man. Immediately the question arises as to who is going to control whom, in whose interest is the controlling going to be exercised, and if control is going to be exercised in the interest of all, who is going to take it upon himself to define that public interest? (Harvey, 1973, p. 125).

To Harvey (1973), Marxist theory provides

> the key to understanding capitalist production from the position of those *not* in control of the means of production . . . an enormous threat to the power structure of the capitalist world (p. 127).

It not only provides an understanding of the origins of the present system, with its many-faceted inequalities, but also propounds alternative practices which would avoid such inequalities:

> we become active participants in the social process. The intellectual task is to identify real choices as they are immanent in an existing situation and to devise ways of validating or invalidating these choices through action (p. 149)

In such a context, geography can no longer be simply academic, isolated in its 'ivory towers'. Its practitioners must become politically aware and active, involved in the creation of a just society which involves replacement of, not reform of, the present one. The remainder of Harvey's (1973) book does not make this commitment clear, however; it contains one essay on land use and land-value theory, investigating the difficult concept of rent, and another on the nature of urbanism, presenting a marxist interpretation of the process of urbanization.

Peet, too, moved from a liberal to a marxist position, replacing his earlier paper on poverty (Peet, 1971) by a marxist interpretation (Peet, 1975a) based on the assumption that inequality is inherent in the capitalist mode of production. This led him to a 'metatheory dealing with the great forces which shape millions of lives' (p. 567) within which

> Environmental, or geographic, theory deals with the mechanisms which perpetuate inequality from the point of view of the individual. It deals with the complex of forces, both stimuli and frictions, which immediately shape the course of a person's life (pp. 567–8).

Such environmental resources act as constraints because they define the milieux within which individuals are socialized and presented their opportunities for participation within the capitalist system. (See also Soja, 1981.) Mere redistribution of income through liberal mechanisms, based on taxation policies, will not solve the problems of poverty, therefore; according to Peet, the requirement is for geographers to seek alternative environmental designs, with removal of central bureaucracies and their replacement by anarchistic models of community control. (Harvey – 1973, p. 93 – disagrees with the latter, pointing out that unless resources are equalized among communities and territories, community control will only result in 'the poor controlling their own poverty while the rich grow more affluent from the fruits of their riches'.)

Structuralism, marxism and realism

As a mode of analysis, marxism is a variant of the philosophy of stucturalism (of which there are many different forms: cf. Johnston, 1983b). In general, structuralism identifies three levels of analysis: (1) the level of appearances, or the *superstructure*; (2) the level of processes, or the *infrastructure*; and (3) the level of imperatives, or the *deep structure*. Of these three, only the first can be directly apprehended; the superstructure of society comprises its social, cultural, political, and spatial organization. But this superstructure cannot be used to account for its own existence. The processes creating it are in the infrastructure. This cannot be observed: its nature can only be theorized and compared with its outputs in the superstructure.

In most forms of structuralism, there is no deterministic relationship between the infrastructure and the superstructure. A particular process can result in a considerable number of outcomes, depending on a variety of factors. To illustrate this, the French social anthropologist Claude Lévi-Strauss developed the analogy of the cam-shaft driving a machine for cutting the outlines of jig-saw puzzles (Figure 6.1). The machine is constrained to produce only certain movements, but the order in which they come is random and means that it can produce a large number of unique puzzles. Study of one of these puzzles alone could not reveal the nature of the machine; only study of the machine could do that. Structuralism is the study of the machines only; but unlike the analogy, these are not available for investigation. The structuralist researcher must develop a theory of the machine, and see whether the contents of the superstructure are consistent

Figure 6.1 A geographical interpretation of Levi-Strauss's cam-shaft analogy for structuralism. Source: Gregory (1978a, p. 100).

with that theory. (Note that the contents of the superstructure cannot be predicted because, to continue the analogy, the particular operation of the machine cannot be predicted.)

In some forms of structuralism, such as Lévi-Strauss's social anthropology (Leach, 1974), it is further argued that the nature of the processes operating in the infrastructure is a consequence of the imperatives of the deep structure. Study of the patterns in the superstructure should reveal the nature of the deep structure, therefore (in somewhat similar fashion to the identification of essences in phenomenology: p. 151). Thus, according to Lévi-Strauss, all incest taboos are variants of a basic taboo which is imprinted in everybody's conscience. That taboo comes out in a slightly different form in each society and culture, but careful comparison of all taboos should reveal their common elements, and thus that aspect of human neural structure.

This search for deep structures does not characterize work in human geography, although it could be claimed that certain spatial concepts – the sense of place; territoriality; orientation – investigated by some humanistic geographers are realizations of deep structures. Only the work of the Swiss structuralist, Piaget, has received much attention. He claims the existence of a deep structure which allows the individual to assimilate and accommodate to new material. Golledge (1981a) argues that this approach informs some of the work in behavioural geography, but there is little evidence that it has had a major impact.

Marxist structuralism for the most part ignores issues relating to deep structures and focuses on the infrastructure and superstructure. Marxism has a materialist base, and so argues that the infrastructure comprises a set of economic processes; most marxist work concentrates on the processes operating within the capitalist mode of production. The contents of the superstructure at any one place in time represent the operations of the infrastructure, within a milieu representing previous operations. They are not determined, however, because the superstructure is created, maintained and changed by human actors. These can interpret the economic processes in a variety of ways, for there is no one way in which a process must be realized – as illustrated in the different party and electoral systems of Western Europe and North America (Taylor and Johnston, 1979). The interpretation must be consistent with the processes, which enable, but also constrain, human decision-making. The results of those interpretations will probably constrain the environment of the future. They will also inform the processes, but will not deflect them.

Marxist analysis seeks to identify the processes operating in the infrastructure and to relate them to the patterns in the superstructure. In human geography, this means deriving general theories of historical materialism that will account for particular patterns. The patterns themselves cannot be used to identify the processes – although they may offer clues as to their contents. The processes can only be identified theoretically, and compared with the superstructure to see if the outputs are consistent with the postulated economic driving forces. Within human geography, for example, Harvey (1975a) has suggested why socio-economic segregation takes place in urban residential areas, as a means of reproducing the class system and defusing inter-class conflict. The basic process is one of distancing (Johnston, 1980b). This can account for the existence of segregation, but not the particular pattern that it takes: suburbanization of the rich is a sufficient but not a necessary response to this process (Walker, 1981a). And for political geography, Taylor (1982) has provided a materialist framework for the operation of the world-economy, within which states act. Again, this provides a theory of what the state is, and what it does, but not of how it reacts to a particular, time-place specific, stimulus (Johnston, 1982a).

Marxist structuralism can also be presented as a variant of the realist philosophy of science. According to this, the natural sciences study the outcomes of processes (the superstructure) but not the processes themselves (the infrastructure). The law of gravity is an empirical regularity, based on repeated experimental observations; this identifies the consequences of an inferred process. In the natural sciences, it is assumed that the processes never change, or that they change in a predictable way which can be inferred from experiments. But in the social sciences built on a materialist base this is not so. The processes themselves are changing continuously (this is the dialectic of capitalism). As a result, the contents of the superstructure are changing too. Empirical laws that are generalizations with universal applicability are not possible in the social sciences – there can be no social scientific equivalent of the law of gravity – because the universe is changing and the same conditions are never repeated.

It is the time-place specificity of the contents of the superstructure which provides one of the major elements in the marxist-structuralist critique of positivism. Positivism, in geography as elsewhere, seeks to identify laws of general applicability; marxists deny that this is possible (Sayer, 1979). It is not quantification which is the basis of the criticism, but its use in the search for positivist goals; quantification as a descriptive tool is perfectly acceptable to structuralists (Walker, 1981b; Taylor, 1981).

Positivist work in human geography is criticized by structuralists, including marxists, because it seeks 'laws' of the superstructure which are unrelated to the processes in the infrastructure, and which in any case cannot exist because of the change that is inherent in the infrastructure. Behavioural and humanistic work is subject to similar criticism, for apparently giving the individual actor complete freedom and therefore removing him from the context in which he operates. With regard to the first, for example, Cox (1981, p. 275) argues that 'behavioral geography . . . unwittingly accepts the assumption of the separation of individual and society'; with regard to the latter, Smith (1979, p. 467) claims that 'The phenomenological approach . . . fails to take seriously the society external to the individual', and Wagner (1976, p. 84) argues that 'Existentialism and its close relatives in the Phenomenological camp seem to abdicate concern with the processes of history and the wider panoramas of geography, and so perhaps lack direct relevance'.

In both positivist and humanistic geography, therefore, the individual is not treated as an actor constrained by the society of which he is a member:

> In 'classical' location theory, agents were individual, autonomous, rational economic automatons. In behavioural theory, the main agents of interest were 'decision-makers' whose prime position is exaggerated by a disregard for economic conditions, thereby producing voluntaristic accounts where 'locational preferences rule' (Sayer, 1982, pp. 80–1)

The structuralist alternative emphasizes that behaviour is constrained by economic processes, but it does not deny human agency. The infrastructure provides 'determinants of activity of which actors are unaware' (p. 81), but within these the individual makes decisions, which in themselves may add to the constraints for future decisions. (Once a spatial structure has been laid down, it influences the form of its own development: Soja, 1981.)

The acceptance of 'knowing actors' operating within structural constraints means that marxism, like other forms of structuralism depicted by the analogy in Figure 6.1, does not argue that the contents of the superstructure are determined by the processes of the infrastructure. Rather it argues that the former can only be understood by the development and 'testing' (Sayer, 1982, pp. 85–7) of theories relating to the latter. Marxism is a theory of the processes. It is not an empirical discipline – its lack of a focus on the details of the superstructure leads Jones (1980, p. 257) to criticize its 'withdrawal from the patterns of distribution which are usually deemed the starting-point of geographical enquiry' – but it provides the framework within which empirical details can be studied and without which their understanding is impossible. However, a marxist geography does accept that the environment itself is an influence on the content of the superstructure. Cultural regions are formed (note that Gregory, 1978a, ends his book with a call for a renewed regional geography focusing on spatial variations in the realization of economic processes). According to Peet (1979, p. 167), in a statement that Duncan and Ley (1982, p. 37) call irresponsible:

> neither environment nor space is passive as social relations pass over and through them. . . . As class relations move over space, they pick up qualities from the regions of a social formation. . . . Class relations become infused with the direct and indirect contents of regions and environments. . . . They are transformed, thereby, into socio-spatial-environmental relations. Spatial relations are at base class relations; class relations contain the effects of space and environment.

Class conflict governs capitalist economic processes. Variations in the interpretation of this conflict from place to place create subtle variations in the superstructure, however. These may become self-sustaining, separate cultures within the overall capitalist framework, with the variations stimulating further spatial differences in actors' interpretations of the ongoing class conflict. (Culture never becomes a determinant, however – Duncan, 1980 – only a secondary constrainer and enabler, secondary that is to the demands of economic processes.) The spatial/environmental dimension is thus important to the marxist analysis of empirical variations (see Quaini, 1982).

Marxism, then, is a form of realism, which seeks to relate the contents of an empirical world to a set of infrastructural determinants – economic processes. It is much more than this, however; it is also the basis for a political ideology aimed at changing those economic processes. According to Peet and Lyons (1981, p. 205)

> If we want to understand what is happening to us, the people of the capitalist world, . . . we need a powerful, logical, sequential, political form of analysis to give an accurate, deep portrayal of the causes of events.

Marxism, including marxist geography, does this, and also provides 'a powerful theoretical and political base for resistance'. The marxist goal is to change the economic processes, and thereby to remove the dominance of the capitalist imperatives on individual action. This goal is based on Marx's humanism. He argued that man is alienated by the capitalist system; that the proletariat is exploited and has its human dignity removed. To restore this dignity, to give the individual full control over himself and his destiny, capitalism must be overthrown and replaced by communism. The argument is that (Relph, 1981b, p. 122):

truly human relationships can be achieved only when everyone can take responsibility for the conditions of their own lives and when there is freedom from the ideologies and actions of a bourgeois professional class.

Capitalism can only be overthrown by revolution, which should be achieved by the proletariat realizing their exploitation and determining to end it. Achievement of that realization requires emancipation, whereby individuals are made increasingly aware of their exploited position. Such achievement is the goal of critical theory (Gregory, 1978a) which combines a marxist realism to the development of communicative skills (hermeneutics) so as to enable individuals to understand the real processes operating in society – to appreciate the infrastructure. Human agency is encouraged, by increasing awareness.

This political goal of marxist work leads to a major element of its critique of positivism – as indicated by Harvey's identification of three types of theory (p. 166). Positivism describes the present situation in the superstructure, as a series of laws. Use of these laws in planning therefore seeks to maintain the present situation. Since that situation is one of exploitation, then positivist planning serves to further exploitation, to retain the present unjust capitalist structure rather than moving it on towards the desired social and economic change. Such work, in any case, rarely understands what is happening in the superstructure, it only describes it. Application of such incomplete understanding is termed instrumentalism (Gregory, 1978a; 1980). Gregory is particularly critical of the use of systems analysis in this context; it provides models that fit the data, and therefore is used to create further fits. Positivist planning seeks to impose its laws on society, rather than recognizing them as instances of a dialectically changing process that never exactly reproduces itself (Marchand, 1978).

To human geographers, therefore, marxist and related work suggests:
1 that explanation of patterns of spatial organization and of man-environment relations within the superstructure can only be understood as realizations of economic processes operating in the infrastructure;
2 that those economic processes cannot be apprehended directly, but only appreciated through the development of theories that are consistent with the outputs in the superstructure;
3 that those economic processes are continually changing, and with them the outputs – as a result universal laws of the superstructure cannot be derived;
4 that class conflict (bourgeoisie v proletariat) is central to the economic processes;
5 that how the processes are realized in the superstructure reflects the actions of individuals operating within the constraints set by the processes and their preceding realizations – individuals are constrained yet enabled, not determined;
6 that any attempt to retain the present superstructure, by planning proce-

dures based on positivist examination of the present, can only help the present unjust system to survive;

7   that understanding the processes and their realizations requires much closer cooperation among, if not integration of, the social sciences than heretofore; and

8   that the goal of academic work must be emancipation, leading to social change.

Adoption of this programme involves altering the model of modern western society. There are two main models, according to Eyles (1974, p. 39):

> one based on consensus, the other on conflict. The first would stress that social life is based on cooperation and reciprocation with norms being the basic element. Such societies would also be characterized by cohesiveness, integration and persistence and life within them would depend on the recognition of legitimate authority, mutual commitments and solidarity. The conflict model sees sectional interests as being the basis of social life which is, therefore, divisive, involving inducement and coercion as well as generating the structural conflicts . . . [that are] the central element in a social system.

Conflict – between classes in a capitalist society – and power – unequally distributed in favour of the bourgeoisie – are the key elements in the latter model, of which marxism is a particular form. Understanding that conflict and how power is distributed and used is basic to understanding how resources are allocated in a society. Its spatial organization is the result of the competition and conflict whose ethic (Eyles, 1964, p. 64)

> implies winners and losers and the winners in economic competition will be those with the powers of ownership and control in the productive process. In this way, growth under such a system will most probably lead to greater inequalities. It would seem obvious, therefore, that poverty and the distribution of real income in a spatial system cannot be understood without reference to power and inequality

Radical theory, research and action

The research tradition generated by the introduction of the marxist (often termed structuralist and/or radical, and sometimes political economy – all are used here) approach has several components. The first is the critique of positivist spatial science and behavioural geography, and of humanistic geography, exemplified by Rieser (1973) and Massey (1975). The second is to provide general theoretical frameworks, within which empirical work can be set (as in Taylor, 1982; Harvey, 1982; and a major collection of essays on urbanization – Dear and Scott, 1981). Thirdly, there is work that seeks to establish how individuals act within the structural imperatives (Johnston, 1983c, illustrates this). Eyles (1981, p. 1386), for example, has argued that although 'no geography or social science can be complete without Marxism', this is insufficient. It must focus on the world of lived experience, the microworld of 'family, network, community, neighbourhood' if it is to be complete, and this requires integrating structuralist marxism with humanistic philosophies. Finally, there is detailed empirical work that seeks to understand particular aspects of the subject matter of human geography within the structuralist framework.

The goal of empirical work in this context is understanding. It seeks to appreciate action, as undertaken by constrained, 'knowing' individuals. It does not seek to explain it, since explanation is a positivist concept implying causation, prediction and general laws. Empirical structural work is carefully phrased to ensure that it is not interpreted as developing generalizations. It is exemplified by the work of Massey and Meegan (1979) on industrial restructuring, who argue that their analyses are historically specific: 'Theoretical argument and empirical evidence alike . . . point to the conclusion that this question – "what is the spatial effects of mergers?" – is not one that can meaningfully be asked' (p. 234). This is because the individual firms involved in the restructuring are not autonomous actors; the strategies available to them are prescribed by the environment in which they are operating, and that environment is always changing (Massey and Meegan, 1982, p. 197). Thus study of the firms themselves cannot provide an explanation for the changes. Similarly, Thrift (1979) shows that the contemporary inner city problem is a particular realization of the problems of capitalist economies, reflecting the special conditions of the times and reactions to them; Pred (1981) illustrates how economic changes under nineteenth-century capitalism affected the use of time resources by families; Archer and Taylor (1981) show how the changing geography of voting can be related to a materialist theory of the American economy and state; and Cox and McCarthy (1982) reinterpret neighbourhood conflict in structuralist terms. Theory and empirical work are to be linked. As Walker and Storper (1981, p. 502) express it:

> Industrial location research and marxist theory have both undergone a revival in the 1970s. It is time they were joined. The task for marxists is to avoid repeating basic principles and to confront the results of empirical research by industrial geographers, for the purpose of developing a better explanatory approach to industrial location.

The same argument is being advanced about most areas of human geography, with work focusing on different combinations of basic theory (macro-scale), major actors (meso-scale), and individual behaviour (micro-scale).

The residential pattern of cities and the operation of housing markets has been a major focus of urban geographers' attention (Herbert and Johnston, 1978). A critique of the positivist approach to the study of those markets was presented by Gray (1975), highlighting:

1 the assumption that people are free to choose where they want to live, when in fact many people are forced to live in particular types of homes in particular places;

2   the assumption that residential patterns result from the amalgamation of a large number of decisions by individual householders, rather than by a few developers and financial institutions; and
3   'there is an implicit belief that the study of individuals and households provides the key to understanding urban processes' (p. 229).
Key words of the positivist paradigm, according to Gray, include aspiration, preference and choice, but:

> There is an accumulating body of evidence from sociologists, and to a lesser extent from geographers . . . that people are not free to choose and prefer from a range of options, and that study of household units does not provide the key to understanding urban processes. Instead, many groups are constricted and constrained from choice and pushed into particular housing situations because of their position in the housing market, and by the individuals and institutions . . . controlling the operation of particular housing systems (p. 230).

To Gray, constraint and competition are more important concepts than are choice and preference, and

> institutions rather than household units are the key to understanding urban processes . . . most fruitfully analysed as reflections of capitalist society. The structural analysis of capitalism and its various spatial manifestations is the core problem facing geographers (p. 232).

While agreeing with Gray that the research emphasizing preference and choice has been one-sided, Hamnett (1977) accused him of throwing the baby out with the bathwater and producing an equally one-sided analysis if he focused only on the production side of housing markets. One cannot study only the constraints and let the choices look after themselves, for many people do have to make choices, within a set of constraints. Gray, he argues, would be better to accept the findings of preference studies and build on them with a structural analysis, thereby combining meso-scale (the allocation of housing resources by manager) with micro-scale (preferences within that allocation) perspectives.

Urban residential patterns have been treated in a series of essays by Harvey, providing a structural analysis. Residential differentiation of various class groups produces community consciousness and the means of reproducing class differences; it is brought about by the actions of finance capital which are aimed at restricting choices and forcing up prices. The result is a series of spatially separate housing markets (Harvey, 1974e); within each, people may move around, expressing their preferences and thereby influencing micro-scale variations in the pattern.

> But there is a scale of action at which the individual loses control of the social conditions of existence . . . [sensing] their own helplessness in the face of forces that do not appear amenable, under given institutions, even to collective political mechanisms of control (Harvey, 1975a, p. 368).

Among the processes at the scale beyond the control of the individual and the community collective is suburbanization, a major means of stimulating consumption – of housing, automobiles, consumer durables etc. – within the American economy at a time of dangerous latent overproduction (Harvey, 1975b, 1978).

Most of the radical work on housing markets and residential patterns in cities has followed Gray's rather than Harvey's point of view, focusing at the meso-scale on the managers who operate the processes involved in the production and allocation of housing. The concept of the 'urban manager' was developed by Pahl (1969), who argued that the residential pattern is the consequence of the operation of two basic processes:

1 There are fundamental *spatial* constraints on access to scarce urban resources and facilities. Such constraints are generally expressed in time/cost distance.
2 There are fundamental *social* constraints on access to scarce urban facilities. These reflect the distribution of power in society and are illustrated by:
bureaucratic rules and procedures
social gatekeepers who help to distribute and control urban resources (p. 215 in 1970 reprint).

The values and ideologies of these gatekeeper/managers are crucial to understanding the socio-spatial pattern of the city, an argument accepted by a group of workers at the University of Cambridge (e.g. Duncan, 1975), whose research included investigations of building-society lending policies (Boddy, 1976), the decisions of local authority housing managers (Gray, 1976), and the role of the local authority in various aspects of the private housing market (Duncan, 1974): similar work has been done in the United States on the operations of real-estate agents (Palm, 1979). Pahl (1975) later revised his ideas, pointing out that planners are merely the 'bailiffs and estate managers of capitalism, with very little power' (p. 7), so that focus on them and on other gatekeepers tends to

view the situation through the eyes of disadvantaged local populations and to attribute more control and responsibility to the local official than, say, local employers or the national government . . . such 'a criticism of local managers of the Caretaking Establishment' and 'of the vested interest and archaic methods of the middle dogs' may lead to an uncritical accommodation to the national elite and society's master institutions (pp. 267–8).

The managers are important elements, but are not themselves the spatial structure of Figure 6.1; they can be interpreted as the cam-shaft, operating the structure. Without care, treatment of the managers outside the constraints in which they operate could become as reductionist as aspects of behavioural geography (see also Williams, 1978, 1982; Leonard, 1982).

Another major area of radical interest has been the modernization of the Third World, in the study of which (Drysdale and Watts, 1977) 'At times, geographers have resembled spectators who, seated on a hill overlooking a battlefield, are fascinated by the direction in which clouds of smoke are blowing' (p. 41). In the positivist mould, modernization had been investigated by Gould and his associates (e.g. Gould, 1970b; Riddell, 1970; Soja, 1968) who selected a series of variables representing various aspects of economic and social change (from a western, technological viewpoint) and manipulated these by multivariate statistical techniques to produce maps of the spread of modernization through colonial territories. Brookfield (1973) accepted that these provided useful descriptions of spatial patterns of change but argued that such 'value-loaded', although ostensibly 'value-free', research reveals little about the process of 'development': he characterizes the work as that of (Brookfield, 1975):

> A superb craftsman ignorant of the material with which he is working. The folly of such an approach was never better demonstrated than in the prolonged failure of . . . [Gould's] group to make its real contribution in an area where the direct participation of geographers is increasingly wanting (p. 116).

Even more worrying to Brookfield is the derivation of policies for development based on notions of diffusion within central-place frameworks. Berry (1972a) produced one of the main papers on this, one which, according to Brookfield (1975) is 'a highly mathematical paper on hierarchical diffusion employing his well-known, but doubtfully relevant data on the diffusion of TV in the United States' (p. 110). (See also Blarkie's, 1978, critique of diffusion studies as a 'spatial cul-de-sac'.) A further critique of the 'geography of modernization' studies (Slater, 1973) led Slater (1975) to identify what he saw as seven weaknesses of the 'Anglo-Saxon mainstream' abstracted empiricism approach:
1   its inverted methodology, which allows data to define problems rather than the reverse;
2   its surfeit of data relative to the power of its theory;
3   its mechanistic abstractions from socio-economic reality, which fail to account for such problems as under-development;
4   its concentration on form-description rather than on process-explanation;
5   its derivative and erroneous theories;
6   its failure to grasp the crucial role of the political economy; and
7   its lack of attention to the class structure and the role of class conflict in the structuring of space.
Brief examples are given of how a theory of uneven spatial development should be written, emphasizing the role of colonialism in the capitalist economy, and thereby moving the focus from the 'sterile perspectives' of the positivist work, highlighting instead that 'A crucial factor in the development of any spatial structure is the way in which surplus is circulated, concentrated and utilized in space' (p. 174). A similar attempt at theory-writing has been made by Folke (1973) with respect to imperialism. (This focus on spatial variations in development has been extended to the regional scale: Dunford, 1982; Massey, 1982.)

Other aspects of geography for which a radical critique has been provided

include Eliot Hurst's (1973) discussion of the impact of highway developments, studies of which have assumed that the socio-political environment
in which the changes took place was itself unchanged. The impact of entrepreneurs on the transport system has been ignored, he claims; statistical procedures cannot account, for example, for the extortion rackets involved in
the construction of California's railway network. For him:

> To really understand network evolution we need to know something about the
> decisions behind its construction. . . . To interpret this material, place it in a
> socio-economic and political framework, overcome the obstacles barring the way
> to the attainment of a better, more humane, more rational social order, and better
> explain and understand transportation as a factor in the landscape, should be our
> ultimate goals in transportation geography (p. 171).

In part, this appears to be a call for an idealist approach (p. 147), except that
Eliot Hurst mentions study of the political and economic frameworks as well
as the actions of the individual decision-makers. The argument, according to
Rimmer (1978), is the third phase in the development of transport
geography, involving a repudiation of the preceding descriptive and
quantitative/predictive phases. Its effect has been:

> to create confusion among transport geographers as the borrowings from socio-
> logy and politics have clashed with those from economics and mathematics. This
> confusion has created strain and tension as transport geographers have tried to
> adapt their ideas to a changing socio-economic environment distinguished by
> uncertainty and turbulence. . . . Transport geographers are, therefore, in a
> dilemma (p. 85).

As a consequence, Rimmer sees a need for a fourth phase, one of redirection,
to produce a humanistic transport geography. Its framework

> has to encompass interactions between individuals and groups in such a way that
> it embraces distributional issues and equity goals in a spatio-temporal frame
> under either advanced capitalism or neocolonialism . . . transport policy has to
> be directed towards minimizing differences in personal mobility and reducing
> the polarization of opportunities produced by different methods of travel
> (pp. 80–1).

Such a redirection clearly does not fit the radical critique.

Work on environmental issues also uses a marxist framework. Harvey
(1974d) has shown that contemporary statements on resource-population
interrelationships are ideological, through an examination of the respective
works of Malthus, Ricardo, and Marx. He argues that 'things' cannot be
understood in isolation from their interactions with other 'things', so that
resources can only be defined in the terms of the relevant mode of production. Thus, predictions such as those in the *Limits of Growth* study (p. 174)
represent a status quo theoretical perspective (p. 166). Other viewpoints
may lead to other conclusions:

> let us consider a simple sentence: 'Overpopulation arises because of the scarcity

of resources available for meeting the subsistence needs of the mass of the population'. If we substitute our definitions [of subsistence, resources and scarcity] into this sentence we get: 'There are too many people in the world because the particular ends we have in view (together with the form of social organization we have) and the materials available in nature, that we have the will and the way to use, are not sufficient to provide us with those things to which we are accustomed' (p. 272).

According to the first version of the sentence, the only policy option available is a reduction in population: according to the second, Harvey identifies three possibilities:
1 changing the desired ends and the social organization;
2 changing technical and cultural appraisals of nature; and
3 changing our views about the capitalist system and its scarcity basis.
The Ehrlich-Commoner arguments (p. 173) are based on the first version of the sentence and, according to Harvey, have been grasped by the elite of the capitalist world as the case for a popular ideology (the need for birth control) which will stabilize the capitalist system currently under severe economic stress. (See also Sandbach, 1980; Sayer, 1981, discusses the analysis of the values which underline the choice of policy options.)

Buchanan (1973) has argued in similar vein, contending that the neo-Malthusian case for birth control in the Third World is part of the 'white north's' imperialist policy of ensuring its continued access to the resources of the international periphery. The problems of ensuring a food supply for an expanding population, plus a supply of needed fuels and minerals, are not insurmountable, he claims, particularly if the residents of the Third World are granted the right to work. 'The poverty which is regarded as symptomatic of reckless population growth is rather a *structural poverty* caused by the irresponsible squandering of world resources by a small handful of nations' (p. 9). The 'white north' is operating an economy of waste, and its theories of overpopulation are grounded in that economy: the apparent humanitarianism of such selfseeking philosophies, is, Buchanan argues, immensely dangerous.

Historical and cultural geography, too, have attracted marxist work. The transition from feudalism to capitalism, and its impress on the landscape, is a topic of increasing interest for the former (Prince, 1979; Peet, 1980; Butlin, 1982), as is the development of the industrial geography of particular regions (Gregory, 1982a). Indeed most areas of human geography have been infiltrated by this approach. To some, this indicates the futility of dividing the social sciences into separate disciplines and sub-disciplines. Eliot Hurst (1980) has called for a de-definition of geography

since the questions and answers that Marxism asks [sic] do not lie within disciplinary contexts as we know them, however much they are changed in response to limited critiques which begin by accepting their pre-existence (p. 3).

To him, geography is 'merely descriptive and scientifically bankrupt . . .

nothing more than a theoretical ideology based on technical practice alone' (p. 7). Marxist geographers do not provide an answer, however, because they maintain the disciplinary separateness, focus on spatial/environmental issues, and prolong the 'fetishism of space' (p. 12). Bourgeois social science reacts to issues with institutional arrangements that are one of: unidisciplinary, multidisciplinary, interdisciplinary, transdisciplinary. None of these is satisfactory. Disciplines must be removed and replaced by the marxist science of historical materialism, which:

> endeavours to discover in its totality all the structural relations which exist between apparent surface appearances and the deeper currents. If one does not do this, then whatever aspect of reality one is attempting to comprehend becomes a fetishized ideological domain (p. 15).

Liberals and radicals in debate

The material discussed in this chapter illustrates very considerable differences of opinion among human geographers during the 1970s, differences which to many are indicative of an ongoing revolution, or paradigm-shift. They form the basis for a debate which, in its published components at least, is much more heated than that generated by the 'quantitative revolution' two decades earlier. It has also been developed in more depth and detail, using forums, such as the journal *Area*, which were not available twenty years earlier, when publication outlets for 'views and opinions' were few. Whereas the behaviourists caused little real concern within the discipline, and their views were soon coopted within the corpus of acceptable approaches, the radicals have had much more impact, because they attack not only the basis of most geographical work but also, in their clear inter-disciplinarity, the bureaucratic structure of the discipline and the existence of geography itself (Johnston, 1978c).

A clear example of the liberal-radical polarization is given by a debate on the geography of crime. This was initiated by Peet (1975b), who argued that in attempting to make their work relevant geographers avoided asking 'relevant to whom?'; the political consequences of their work were ignored. The studies of crime reviewed (e.g. Harries, 1974) refer only to the surface manifestations of a social problem; they cannot provide solutions to that problem, but only ways of ameliorating it. 'So it is that "useful" geography comes to be of use only in preserving the existing order of things by diverting attention away from the deepest causes of social problems and towards the details of effect' (p. 277). Furthermore, geographers study only the crimes for which statistics are collected, thereby accepting the definition of crime by the elite; the maps which they produce, which are useful to police patrols, can therefore be employed to help maintain the status quo of power relations within society. The position of the geographers involved, a position which they do not declare, is one of protecting the 'monopoly-capitalist' state.

Harries (1975) responded by attacking the simplistic nature of Peet's arguments, and claimed that geographers would have no influence at all if they argued merely that crime is a consequence of monopoly capitalism. He argued that it is best to work within the system, to make the administration of justice more humane and equitable, to protect the potential victims of crime, and to provide employment opportunities for graduate geographers. Approaches based on the control of crime are likely to be more influential than polemics relating its cause to the mode of production: as Lee (1975) also pointed out, Peet 'failed to provide us with any clues as to how he or other radical geographers would study crime' (p. 285).

Peet (1976a) responded to Lee's challenge, presenting a radical theory which would 'contribute directly, through persuasion, to the movement for social revolution' (p. 97). The theory argues that capitalism harnesses human competitive emotions and produces inequalities of material and power rewards. Aggression is an acceptable part of being competitive, and is often released on the lower classes, who are encouraged to consume but provided with insufficient purchasing power. As the contradictions of this paradox increase, so does the pressure to turn to crime. Thus crime occurs where the lower classes live, and at the spatial interface between them and the middle class. Harries (1976) replied that being a radical was a luxury few academics could afford; working at a publicly-financed university demanded a pragmatic rather than a revolutionary approach unless unemployment was desired. To him, Peet's theory is overly economic and deterministic; it fails to account for cultural elements, such as the disproportionate criminal involvement of blacks, the sub-culture of violence in the southern United States and other areas, and the fact that all economic systems produce minorities disadvantaged in terms of what they want and what they can get by socially legitimate means. He could offer no alternative theory, however: 'I do not carry in my head a theory of crime causation, and I am quite incapable of synthesizing and attaching value judgements to existing theoretical formulations within a couple of pages of typescript' (p. 102). He encouraged Peet to come off the fence, and to get involved in the production of change within the present system. Wolf (1976), on the other hand, claimed that by concentrating on the traditional concerns of marxism Peet was not radical enough.

Two geographers who have been involved in a considerable, often virulent, debate are Brian Berry and David Harvey. Berry (1972b) initiated the exchanges with his comments on Harvey's (1972) paper on revolutionary theory and the ghetto. Berry wondered whether Harvey's rational arguments on the need for a revolution would be accepted: 'because of "commitment", the opposition will quietly drift into corners, the world will welcome the new Messiah, and social change will somehow, magically, transpire' (p. 32). The power to achieve change, he contended, needs more than logical argument to produce it in the twentieth century – 'nothing less than cudgels has been effective' (p. 32) – and Harvey's belief in logical rationalism will be to no avail. He has also argued that, in any case, Harvey is wrong about the ghetto, for liberal policies are succeeding and the inequalities between blacks and whites are being reduced (Berry, 1974a). Harvey (1974a) responded that scarcity must continue in a capitalist economy, which will leave some people – those in the inner city – relatively disadvantaged.

In reviewing *Social Justice and the City* (Harvey, 1973), Berry (1974b) criticized Harvey's dependence on economic explanations. Basing his case on the arguments of Daniel Bell (1973) on post-industrial society, Berry (1974b) claimed that the economic function is now subordinate to the political: 'the autonomy of the economic order (and the power of the men who run it) is coming to an end, and new and varied, but different, control systems are emerging. In sum, the control of society is no longer primarily economic but political' (p. 144). (His comments clearly pre-date the economic recession of the mid and late 1970s.) Harvey's (1974b) response was that marxism could not be considered as passée while the selling of labour power and the collusion between the economically and the politically powerful continue, and that the state has to be considered too within a marxist framework (Harvey, 1976). Berry (1974b) retorted:

> I believe that change can be produced *within* 'the system'. Harvey believes that it will come from sources *external* to that system, and then only if enough noise is produced at the wailing wall. . . . The choice, after all, is not that hard: between pragmatic pursuit of what is attainable and revolutionary romanticism, between realism and the heady perfumes of flower power (p. 148).

Harvey's (1975c) review of Berry's (1973a) *The Human Consequences of Urbanization* – a study of urbanization processes at various times and places and of the planning response to these – concluded that the book is 'all fanfare and no substance' (p. 99), revealing that Anglo-American urban theory is substantively bankrupt and that 'it is scholarship of the Brian Berry sort which typically produces such messes' (p. 99).

> It is doubtful if it makes any sense even to consider urbanization as something isolated from processes of capital formation, foreign and domestic trade, international money flows, and the like, for in a fundamental sense urbanization is economic growth and capital accumulation – and the latter processes are clearly global in their compass (p. 102).

To Harvey, Berry has nothing to say of any substance, but as Berry 'is influential and important . . . his influence is potentially devastating' (p. 103). Berry's only response was a general comment on the Union of Socialist Geographers (Halvorson and Stave, 1978):

> there's no more amusing thing than goading a series of malcontents and kooks and freaks and dropouts and so on, which is after all what that group mainly consists of. There are very few scholars in the group (p. 233).

Apart from these very polarized exchanges, there have been several other statements which indicate that whereas some 'liberals' have been prepared to consider the radical case seriously, others have tended to avoid the issues, more or less. Chisholm, for example, has (1975) claimed that:

> while I am fully sympathetic to the view that the 'scientific' paradigm is not adequate to all our needs, and must be supplemented by other approaches, I am not persuaded that it should be replaced. . . . Harvey wants us to embrace the marxist method of 'dialectic'. This 'method' passes my understanding; so far as it has a value, it seems to be as a metaphysical belief system and not — as its protagonists proclaim — a mode of rational argument (p. 175)

(see also Chisholm, 1976; Sayer, 1981). More frequently, reviewers accept that the radical view is valuable but not, to them, tenable in its entirety. Thus Morrill (1974) writes of Harvey's *Social Justice and the City* that 'I am pulled most of the way by this revolutionary analysis but I cannot make the final leap that our task is no longer to find truth, but to create and accept a particular truth' (p. 477). King (1976), too, seeks a middle course,

> An economic and urban geography that will be concerned explicitly with social change and policy. . . . Such a middle course will not find favour with the ideologues, who will see it either as another obfuscation favouring only 'status quo' and 'counter-revolutionary' theory, or as a distraction from the immediate task of building elegant quantitative-theoretic structures, but some paths are being cut through the thicket of competing epistemologies, rambling lines of empirical analysis, and gnarled branches of applied studies that now cover the middle ground (pp. 294–5).

He accepts that much quantitative-cum-theoretical geography has sought mathematical elegance as an end in itself, at the sacrifice of realism; he believes that social science must feed into social policy and generate social change; he accepts the 'intellectual power' of marxist analysis but believes that the prescriptions based on it are acceptable only if the ideological framework is: his conclusion suggests the need for more quantification, aimed at being operationally useful rather than mathematically elegant. He later notes that '*space* . . . should be seen as an element in the political process, an object of competition and conflict between interest groups and different classes' (King and Clark, 1978, p. 12). And finally, Smith (1977) concluded that:

> Marx may have been able to dissect the operation of a capitalist economy with particular clarity, and see the essential unity of economy, polity and society that we so often miss today. But Marx does not hold the key to every modern problem in complex, pluralistic society (p. 368).

The debates continue, very largely it seems because many geographers are unable/unwilling to accept both the marxist analysis of society and the marxist programme for action. In part, this reflects a partial reading of marxism – in particular a concentration on strict structural interpretations which do not apparently allow for the activities of knowing individuals and which do not accept the concept of structuration (see p. 196). Thus Ley (1980, p. 12) identifies in marxism 'some hidden transcendental pheno- menon . . . directing the course of human society': this commits an epistemological error, denying or at least suppressing the subjective; a theoretical error that devalues the power of human action to redirect the course of events; and a moral error, which makes humans into puppets and threatens basic freedoms of speech, assembly and worship. Muir (1978) attacks interpretations of marxism, and similarly implies that it threatens individual freedom of the academic 'to pick and choose from among the . . . literature' (p. 325). Respondents point to the lack of such 'intellectual ortho- doxy' and 'sterilized geography' (Manion and Whitelegg, 1979; Duncan, 1979), but Muir (1979, p. 127) remains convinced of the threat implicit in 'the commands from such little men as Marx, Lenin, Trotsky, Stalin and Mao concerning the primacy of activism, the obligations of party member- ship and the necessity to subordinate individual judgement to the will of the party'. Radical geography is contributing to understanding, but to call it marxist is to give it a certain programmatic base. To Walmsley and Sorensen (1980) marxism is just irrelevant, and deflects attention from 'reformism and relevance'.

Duncan and Ley (1982) have published an extensive critique of structural marxism in geography. This contains four major themes. First, marxist analysis is a form of holism, in which the whole – variously termed capital, the economic structure, economic processes, etc. – is given a life of its own: an abstraction is assumed to exist. Such an act (a reification) offends their belief in man as a conscious, free agent (see Duncan, 1980, on holism in cultural geography, and Agnew and Duncan, 1981) – although they do not proclaim an idealist alternative and accept that 'individual action cannot be fully explained without reference to the contexts under which individuals act' (p. 32). Secondly, individuals are represented as agents of the whole – the means of implementing its goals – not free decision-makers in their own right. Thirdly, the materialist infrastructure of marxism is a form of economism, in that it presents economic processes as the ultimate cause of all behaviour. This excludes, to them, many other influences and leads to the final theme, 'the attempt to cast explanation continually and everywhere in terms of economic imperatives, leading . . . to a crisis in empirical exposi- tion' (p. 47). Here, they make what is essentially a positivist critique:

> the form of the explanations is both tautological and empirically untestable. The result is a mystification in explanation of how real processes operate (p. 55).

They conclude that structural marxism in human geography presents a passive view of man, and offers explanations in terms of abstract wholes which are obfuscatory and not verifiable.

Structural marxism is criticized by others who disagree with its treatment of the individual, and who wish to build more convincing models of the interrelationships between infrastructure and superstructure in the process of societal change. Gregory (1981) presents four such models (Figure 6.2):

1 Reification, in which the individual is entirely determined in his action by the whole – as in a structural marxism that denies human agency.

2 Voluntarism, in which society has no separate identity but is constructed from the actions of free individuals.

3 Dialectical reproduction, in which the whole creates the individuals, whose actions influence the whole – which in turn creates the next generation of individuals.

4 Structuration, which starts with individuals (not with the whole) and

A   REIFICATION   typified in social theory by Emile Durkheim and by some neo-Marxist formulations

SOCIETY

INDIVIDUAL

Society is a reality *sui generis* which is external to and constraining upon human agency

B   VOLUNTARISM   typified in social theory by Max Weber

SOCIETY

INDIVIDUAL

Society is constituted by intentional action

C   DIALECTICAL REPRODUCTION   typified in social theory by Peter Berger

SOCIETY   SOCIETY

INDIVIDUAL   INDIVIDUAL

Society forms the individuals who create society in a continuous dialectic: society is an externalization of man, and man a conscious appropriation of society

D   STRUCTURATION   typified in social theory by Jürgen Habermas and Anthony Giddens

SOCIETY ——————— SOCIETY

INDIVIDUAL ——————— INDIVIDUAL

Social systems are both the medium and the outcome of the practices that constitute them: the two are recursively separated and recombined.

Figure 6.2   Four models of the relationships between society and the individual. Source: Gregory (1981, p. 11).

portrays individuals and society in a continuous dialectic from then on. The last of these is preferred by Gregory, because it offers a full integration of humanist and materialist (or structuralist) perspectives – in a way that critiques such as those of Duncan and Ley do not. Such an integration, he claims, will recognize the materialist base to society – the infrastructure – while accepting the important role of human agency. Humanistic geography should not compete with a scientific approach, but should acknowledge:

> the recurrent and recursive relations between the individual and society as being fundamentally implicated in the production and reproduction of both social life and social structure (p. 15).

Alongside these debates on the relevance of marxist ideas, other geographers have been defending quantitative spatial science against critiques such as Gregory's (1978, 1980). Bennett (1981d, p. 24), for example, argues that

> much of the critique of quantitative geography as 'positivism' has been misplaced and has accepted uncritically the representation of science as positivism given by Harvey in 1969. This has had a pernicious and destructive effect on the subject in three main ways: first, it has suggested that scientific and empirical enquiry is largely socially worthless; second, it has often rejected the links between physical and human geography; and third, it has often rejected the existence of geography as a discipline at all.

(Elsewhere, he has characterized the critique as 'at best a mis-representative irrelevance, and at worst a fatuous distraction': Bennett and Wrigley, 1981, p. 10.) The argument appears to be the valid one that quantification need not be allied with positivism. Bennett identifies three messages from Harvey's *Explanation in Geography*: geography is primarily inductive, is an objective science, and seeks universal laws. But 'each is only a partial representation of the literature and ideas it seeks to describe' (Bennett, 1981d, p. 13). He argues that radical and quantitative approaches must be integrated:

> by empirical analyses, from the integration of environmental, social, political, and economic aspects with space, historical stimuli, and specific modes of thought and their spatial-political manifestations, . . . by the re-establishment of the geographical subject matter (p. 24).

He wishes to maintain a separate discipline of geography (human plus physical), in which quantitative geography – 'never . . . truly positive' (Bennett and Wrigley, 1981, p. 10) – occupies a central place, answering 'the fundamental questions of social norms, social distribution, policy impacts and humanistic concerns which the critics rightly emphasize' (p. 10). To many 'radicals', this would appear to be a *status quo* approach (p. 166).

Other writers have sought to integrate elements of the positivist, humanistic and structuralist approaches to human geography (e.g. Christensen, 1982). Some attempts have been short-lived, such as an essay on the links between catastrophe theory (p. 106) and the discontinuities central to marxist economic theory (Day and Tivers, 1979, 54–8; Alexander, 1979, 228–30). Hay (1979a, p. 22) has argued for an empirical, analytical geography that is:

> at the same time a *nomological* geography which seeks, for example, to understand the workings of urban rent theory as positivistically observed, a *hermeneutic* geography which seeks to identify the meaning of the urban rent system for those who are participants (active or passive) within it, and a *critical* geography which points to the extent to which present urban rent systems are themselves transformations of the capitalist system, but which admits that some of its features may indeed be 'invariant regularities'.

Livingstone and Harrison (1981, p. 370) somewhat similarly present the case for:

> a humanistic geography which is, at the same time, critical, in questioning rather than bracketing our presuppositions, hermeneutic, in interpreting the meanings behind action, and empirical, in examining the subjectively interpreted objective world.

They, it should be noted, include no nomological – or generalizing – component.

To some, particularly the critics of positivism, such an attempted integration is not possible because the approaches are incompatible: Gregory (1978a, p. 169) terms it 'inchoate eclecticism' (see also Gregory, 1982b; Eyles and Lee, 1982. Positivist and structuralist approaches sit unhappily together in Rhind and Hudson, 1981.) The biggest problem is with positivism – not quantification (Walker, 1981b) – because of its nomological orientation and its belief that human geographers can discover 'invariant regularities'. (This avoids the issue of the length of a time-period. According to marxist theory, the infrastructural processes are always changing. If the change is slow, however, there may be periods in which virtually invariant regularities can be observed: Johnston, 1980a. The problem arises if these are treated as entirely invariant, and used in predictive and planning exercises to bolster the status quo.) At present, the major links are between the structuralist and the humanistic approaches (as suggested in Eyles, 1981), which treat the knowing actor operating within structural constraints. Most of the humanistic approaches, as they have been presented to human geographers, focus entirely on the human as a free decision-maker, with no reference to either structural constraints or to how an individual's decision-making capabilities develop in a societal context. But, as several authors have indicated (e.g. S.S. Duncan, 1981), human geography as an empirical discipline must marry the ability to develop realist theories of the infrastructure with methods of studying human action within the super-structure (Johnston, 1983c). Such marrying represents a positive critique, rather than the almost entirely negative debate reviewed earlier in this section.

Conclusions

This chapter has brought the debate on the nature of human geography up to date at the time of writing, and has indicated the depth and breadth of the discussions which have occupied much of the 1970s. The clearest conclusion which can be drawn is that there is an increasing proportion of geographers who wish to be involved in the re-shaping of societies, either through ameliorative correction of current problems and trends or by designing desirable spatial organizations (Berry, 1973a): to some, this is necessary if human geography in to retain its institutional position. Their motives range along the continuum from 'pure altruism' to 'devoted self-seeking'. Their methods vary from those who accept the present mode of production and see humanitarian goals as achievable within its constraints, through those who subscribe to the phenomenological view (Buttimer, 1974) that:

> the social scientist's role is neither to choose or decide for people, nor even to formulate the alternatives for choice but rather, through the models of his discipline, to enlarge their horizons of consciousness to the point where both the articulation of alternatives and the choice of direction could be theirs (p. 29),

and ending with those who believe that a revolution is necessary to remove the causes of society's myriad problems and replace them by an equitable social structure.

Perhaps somewhat surprisingly, the 'revolutionaries' of the above classification are not the current 'activists'. It is the liberals who believe that geographers can contribute to the immediate solution (or amelioration) of current problems and who are pressing for academic involvement in policy-making: the radicals have a longer-term goal of education which will generate a mass movement demanding revolutionary change. The two groups are members of the same discipline. Contact between them is in many cases either virulent or non-existent, it would seem from their published exchanges. And alongside them are those little affected by the debate, whose interests, be they in historical, cultural or regional geography, lead them to believe that their task is to continue as before, sensitizing their readers to the world as they, the geographers, perceive it.

Currently, therefore, Anglo-American geography is characterized by a plurality of approaches (Gould and Olsson, 1982). These can be classified into three philosophies, each with a separate epistemology, or theory of knowledge. First there is the positivist philosophy, with its belief in the objectivity of scientific description (empiricism) and analysis of the world, its goal of formulating laws about that world, and its assumption that explanation (causal laws) can be derived by studying the outcomes of the laws; the laws of spatial organization can be revealed by analysing spatial patterns. Secondly, there are the humanistic philosophies based on a belief that men live in subjective worlds of their own creation, within which they act as free agents. Their actions cannot be explained (predicted) as examples of general laws of behaviour, but only understood – or appreciated – through methodologies that pierce their subjectivity. Finally there are the structuralist, or realist, philosophies which argue that explanations for observed patterns cannot be discovered in the analysis of those patterns themselves, but only by the development of theories of the underlying – although inapprehendable – processes that generate the conditions within which human agents can create those patterns. Foremost within this group is marxism (particularly humanistic marxism) which argues that the processes are themselves changing – and can be altered by concerted political action – so that no laws of spatial organization are possible. In addition, there are still arguments for regional geography (Hart, 1982.) For human geographers, this plurality is a focus of debate, and even a source of confusion. For some, it presents a polarization that can only be solved when one (paradigm?) is proved triumphant. (Others argue that this cannot be done logically, since there are no common criteria for comparing the approaches.) For many, the plurality offers potential for developing a newer, more robust human geography, though the nature of that development is as yet far from clear.

7

Evaluation

The previous five chapters have outlined the major debates that have taken place within Anglo-American human geography since 1945 with regard to the discipline's contents – what it studies, how, and why. The chapter titles themselves indicate that several very different approaches to the discipline have been advocated. (For more detail on the approaches themselves, and on the substantial applications, see Johnston, 1983b.) The purpose of this final chapter is not to assess progress within those approaches, let alone to establish whether they have contributed towards the attainment of 'higher levels of intellectual, social and physical well-being for [our] fellow men' (Wise, 1977, p. 10). Rather, the evaluation here returns to the issues raised in Chapter 1, where several different models of the development of scientific disciplines were presented. No formal testing of those models is presented, for no methodology has been outlined that would allow such a task. Instead the general relevance of the ideas of Kuhn, Lakatos, Popper, Mulkay and Foucault is assessed against the material outlined above.

Human geographers and models of disciplinary progress

Along with members of almost every other academic discipline, human geographers have been attracted to the ideas and language of Kuhn's paradigm model. As Harvey and Holly (1981, p. 11) express it:

> the use of the word paradigm has become fashionable in geography as well has having become a pivotal concept for courses in geographic thought on both sides of the Atlantic. Thomas Kuhn has become as familiar to students of geography as Hartshorne or Humboldt!

In general, the use of Kuhnian ideas has been undertaken with relatively little research into the major debates that they have stimulated throughout Anglo-American academia. Most human geographers rely on the first, 1962, edition of *The Structure of Scientific Revolutions*; they seem to be unaware either that 'The loose use of "paradigm" in his book has made [it] amenable to a wide variety of incompatible interpretations' (Suppe, 1977a, p. 137) or that 'Kuhn's views have undergone a sharply declining influence on contemporary philosophy of science' (Suppe, 1977a, p. 647). Kuhn (1977) himself has substantially revised his ideas. (See also the exegesis of this work in Barnes, 1982.) Agnew and Duncan (1981, p. 42), for example, have argued that

> Recent reviews and programmatic statements concerning trends in Anglo-American human geography leave the impression that little attention has been given by geographers to the philosophical compatibility of borrowed ideas . . ., that the political implications of different ideas have largely been ignored . . ., and that controversy on source disciplines or literatures has not excited much interest.

Their examples do not include the import of Kuhnian ideas into geography, but their conclusions certainly hold in this case too.

Kuhnian concepts were first used in the geographical literature by Haggett and Chorley (1967), in a normative sense as part of an argument for a revolution in geographical method. They identified an inability of the then dominant paradigm (discussed in Chapter 2 of the present book) to handle both the explosion of relevant data for geographical research and the increasing fragmentation and compartmentalization of the sciences. They proposed a new 'model-based' paradigm

> able to rise above this flood-tide of information and push out confidently and rapidly into new data-territories. It must possess the scientific habit of seeking for relevant pattern and order in information, and the related ability to rapidly discard irrelevant information (Haggett and Chorley, 1967, p. 38)

The launch of the new paradigm had occurred more than a decade previously (see Chapter 3 above); Haggett and Chorley's goal was to diffuse the new ideas, and to win British converts to a new disciplinary matrix, if not world view, for study in human geography.

It was the concept of scientific revolutions that attracted Haggett and Chorley. A very similar concept was argued by Burton (1963), whose paper appeared almost contemporaneously with Kuhn's book, and contained no reference to it. Buton introduced the term 'quantitative and theoretical revolution', and argued that

> *An intellectual revolution* is over when accepted ideas have been overthrown or have been modified to include new ideas. *An intellectual revolution* is over when the revolutionary ideas themselves become part of the conventional wisdom. When Ackerman, Hartshorne and Spate are in substantial agreement about something, then we are talking about the conventional wisdom. Hence, my belief that the quantitative revolution is over and has been for some time. Further evidence may be found in the rate at which schools of geography in North America are adding courses in quantitative methods to their requirements for graduate degrees (Burton, 1963, p. 153)

Similarly, Davies (1972b) used Kuhnian terminology in the title of his book, *The Conceptual Revolution in Geography*. The revolution that he identified involves a change from the contemplation of the unique to the adoption of 'the more rational scientific methodology' (p. 9): none of the contributions to his book refer to Kuhn, but the whole is clearly set in a context strongly influenced by Kuhnian ideas. And a little later, Harvey (1973) used the term, like Haggett and Chorley, in a normative sense in his search for a new world view.

The authors cited so far have used the paradigm concept at either the macro-scale of a world view or the meso-scale of a disciplinary matrix (see above, p. 23). Others have used it as a general descriptive tool (e.g. Buttimer, 1978b, 1971; Holt-Jensen, 1981; it has the largest number of entries in the index to James and Martin, 1981); and as a framework for summarizing sub-disciplinary changes (Herbert and Johnston, 1978). And yet others have used Kuhn's micro-scale concept of a paradigm as an exemplar. Taylor (1976), for example, identified seven separate revolutions during the period under review here and Webber (1977) proposed an entropy-based paradigm (see p. 121). Harvey and Holly (1981) have made explicit use of the exemplar concept. They identify five paradigms within geography during the last century, associating each with an individual scholar

> we can tentatively assign paradigmatic status to . . . Ratzel with the paradigm of determinism, Vidal with that of possibilism, Sauer with the landscape paradigm, Hartshorne with the chronological paradigm and Schaefer with the spatial organization paradigm (Harvey and Holly, 1981, p. 31)

In effect, they are identifying 'schools of thought' associated with particular scholars, some of which existed concurrently rather than consecutively. In the 1960s, they claim, a single paradigm dominated (whether its leader was Schaefer is open to question, despite the views of Bunge, 1979, and others), but the 1970s was characterized by a 'diversity of viewpoints' (p. 37), within which spatial organization remains important. Zelinsky (1978, p. 8) calls the 1970s

> a decade of confused calm, or rather of pluralistic stalemate, as geographers explore a multiplicity of philosophical avenues and research strategies . . . without that single firm conviction as to destination that guided most of us in the past.

The discipline has become technically more capable, substantively more catholic, philosophically more mature, internationally more merged, socially more relevant, and academically more linked to other disciplines, he claims. And this will probably lead to a continued plurality of approaches:

> I happen to believe that, more than anything else, this philosophical coming of age, this rising above the superficiality and tunnel vision that blemished so much geographical work earlier in this century, justifies (p. 10)

the title 'Human geography: coming of age'.

Although some are uncertain about the relevance of Kuhn's concepts to the 1970s and 1980s, in comparison with the 1950s and 1960s, others are less equivocal. In his introduction to a series of essays on *The Nature of Change in Geographical Ideas*, Berry (1978a) claims that

The changes in geographical ideas that we have discussed are distinctly Kuhnian
. . . What, then, is progress in geography? The perspective provided by the
essays in this book is distinctly Kuhnian (pp. vii, ix).

There is little evidence of this in the essays by other contributors, however,
and Berry's own contribution suggests that pluralism and inter-paradigm
conflict have been much more common than have periods of normal science.
Writing of geographical theories of social change, he notes (Berry, 1978b,
pp. 19, 22):

a diversity arising from the mosaic quality of modern geography . . . they have
. . . moved from one paradigm to another, and in the last decade they have been
extremely dynamic. . . . With multiple ideas and multiple origins, modern
geography could rightly be characterized as a 'mosaic within a mosaic' (Mikesell,
1969).

Human geographers and a critique of Kuhnian applications
Kuhnian concepts and terminology are widely used in recent human geography literature, therefore, although it should be noted that some recent treatments of disciplinary history entirely ignore this literature (Freeman, 1980a,
1980b). Many of the presentations are superficial, however (see Graves,
1981), and are both cavalier and uncritical in basing their descriptions on
Kuhnian foundations.

But not all geographers are so convinced of the value of a Kuhnian
interpretation, with two authors going so far as to claim that this has
'distorted even perverted the development of geography' (Young and Petch,
1978, p. 1). The concepts of revolutions and of normal science have both
been criticized as poor descriptions of geography in recent decades. Individual geographers may have experienced personal revolutions and rapid
shifts from one paradigm (at the meso- if not the macro-scale) to another
(Harvey, 1973, indicates this for himself: see the essays in Billinge, Gregory
and Martin, 1982). But to claim that such experiences can be amalgamated
into disciplinary revolutions, of which a number have been identified,
strikes some observers as both inapt and inconsistent with the evidence:

Perhaps so many revolutions in so short a time indicate in themselves either a
continuously rolling programme, or something basically wrong with the over-
turning metaphor (Bird, 1977, p. 105)

The reason for this, Bird (1978, p. 134) suggests, is that mono-paradigm
dominance of a discipline is inconsistent with 'the fact that society itself is
organized around more than than one major principle'.

More pointed criticism is made by Stoddart, who initially saw some value
in the paradigm model (Stoddart, 1967b). He argues (Stoddart, 1977, p. 1)
that

the concept sheds no light on the processes of scientific change, and readily
becomes caricature. I suggest that as more is understood of the complexities of

change in geography over the last hundred years, and especially of the subtle interrelationships of geographers themselves, the less appropriate the concept of the paradigm becomes.

His case is based on an analysis which shows both the absence of consensus (normal science) and the slow pace of change (which is more readily represented in Lakatos's schema). The concept of a paradigm, he claims, has become part of the 'boosterism' image with which geographers conduct debates. Thus (Stoddart, 1981a):

> the paradigm terminology has been used to illuminate either the establishment of views of which a commentator approved, or to advocate the rejection of those he did not (p. 72) . . . the concept of revolution bolsters the heroic self-image of those who see themselves as innovators and who use the term paradigm in a polemical manner . . . those who propound the Kuhnian interpretation have done so in ways which tend to make it self-fulfilling (p. 78)

and (Stoddart, 1977, p. 2):

> There is scope for sociological enquiry into the extent to which the concept has been used in recent years as a slogan in interactions between different age-groups, schools of thought, and centres of learning

Despite such criticisms, little has been done to suggest an alternative sociology, either based in another model of the history of science or developed specifically to represent the situation in human geography. Wheeler (1982) finds Lakatos's work attractive, arguing that several separate programme cores can be identified in contemporary human geography (he cites areal differentiation, spatial science, cognitive-behavioural approaches and marxist structuralism), in each of which the operation of positive heuristics (see p. 24) can be identified. These research programmes are in continuing competition, whose nature will change over time (Wheeler, 1982, p. 4):

> Given the deficiencies in Kuhn's scheme and the rather ill-defined nature of geography, it seems unlikely that the future of the discipline will be characterized by sequences of revolutionary change interspersed with efficient problem solving. Moreover expectations of revolutionary progress appear to be unjustified. Instead it seems that a variety of approaches, which will undoubtedly wax and wane in popularity, will continue to be employed.

Geography and its environment

A thesis developed recently by some geographers is that the major influence on the content of the discipline is its environment, particularly its economic, social and political milieux (the contents of which may influence other disciplines more directly, which than transmit changes to human geography). Stoddart (1981b, p. 1), for example, introduces a book of essays which demonstrates

> that both the ideas and the structure of the subject have developed in response to complex social, economic, ideological and intellectual stimuli.

Evidence for this is presented, as a second-theme of the essays, in the reciprocal relationship between geographers and their milieux:

> thoughout its recent history geographers have been not only concerned with narrowly academic issues, but have also been deeply involved with matters of social concern.

A framework for such a contextual approach, which emphasizes neither external factors nor those internal to the discipline, has been outlined by Berdoulay (1981). It has been illustrated by Grano (1981, p. 30):

> It was the external goals of society that brought the establishment of geography as an academic discipline. This took place without any notable contribution from any other scientists.

Geography was established, and institutionalized, because it was seen to be beneficial to society. (In Grano's example, the benefits flowed to the Prussian state. Freeman, 1961, 1980a, illustrates a similar theme for British geography. See also Reiser's, 1973, linking of changes in the discipline's content with changes in its capitalist environment.) Once established, geographers created their own guidelines and disciplinary identity, but as the milieux changed, so did the guidelines and the identity. As Capel (1981, p. 66) argues:

> The established community employs strategies tending to reproduce and amplify itself. Never will it opt for self-liquidation; the community will defend its survival, even if other communities of scientists investigate similar problems with like method, or if the logical incoherence of the conceptions that they defend is revealed. . . . Everything will be sacrificed for the reproduction and growth of the community, including the coherence of the very conception of the discipline: different conceptions can defend themselves in distinct moments or even simultaneously, without putting into doubt the continuity of the science practised.

If this is the case, then the study of the history of human geography should be phrased as the identification of separate discourses set in a variety of milieux (Claval, 1981; see also Scott, 1982).

This 'environmental influence' presentation of the history of a discipline has been employed by Mikesell (1981) in a description of the history of geography in the United States. That history has been characterized, he claims, by a sequence of 'temporary enthusiasms or episodes' (p. 9). The content of these episodes is a response to the stimuli of the contemporary American environment – hence the popularity of urban studies in recent decades, the decline in foreign-area studies after the Viet-Nam withdrawal, and the impact of national concern about the physical environment in the 1970s. Furthermore, he also suggests the role of environmental influences on the culture of individual geographers:

> The first generation of American geographers grew up in a country that was still strongly influenced by the *mores* of small towns. Most of the students now

attracted to geography are products not only of an urban but increasingly of a suburban environment . . . the geographical profession has changed and is changing as a consequence of the suburban, middle-class origin of most of its current members (pp. 11–12).

Porter (1978) has advanced similar arguments, identifying two types of American geography: a Midwest version that 'was a characteristically optimistic, action-oriented, "can-do" kind' (p. 17); and a Californian geography, based on Sauer, 'historical, uncompromisingly academic, speculative, suspicious of government, keenly interested in cultures other than the dominant Anglo culture of the United States' (p. 18).

This environmental influence thesis can be taken too far, of course, and should not be allowed to become environmental determinism: Mikesell notes that urban geography flowered in the 1950s and 1960s not only in Chicago but also in Iowa. But it does suggest that human geography and human geographers must be studied in the context of their milieux, and that, as Capel (1981) so clearly points out, the disciplinary community (or major elements within it) will seek to maintain its identity by bending to the perceived environmental shifts. Further, as Mikesell (1981, p. 13) argues – following Harvey (1973) and others – this reaction to environmental shifts is undertaken by individual scholars, who are seeking not only to defend and promote their chosen discipline but also to defend and promote their own status and careers within it:

> innovation will continue to be regarded as a virtue. Much of the development that has already taken place in American geography is a consequence of the attempt of individual scholars to stake a claim for themselves, to be or at least to seem to be different from their rivals. The fact that most academics see virtue in innovation means that there is reward for innovation.

What Mikesell does not discuss is the scale of the innovation; does it have to involve developments within the positive heuristic of a research programme, to use Lakatos's terms, or does it require the launch of a new research programme? Can one bring potentially greater rewards, and disasters, than the other?

The relevance of the environmental influence thesis has been illustrated several times in this book. It has been especially relevant in the late 1970s and early 1980s as a consequence of the economic recession, the cut-back in higher education, and the attacks on social science research (see p. 35). Many have reacted to this by promoting human geography as an 'applied discipline', offering relevant skills for the attack on contemporary problems. (This is illustrated in Chapter 6.) Writing in the December 1981 issue of the *Newsletter* of the Association of American Geographers (AAG), under the title 'A survival package for geography and other endangered disciplines', Kish and Ward (1981, p. 8) state that:

> On 19 June, 1981, the Board of Regents of the University of Michigan decided, by unanimous vote, to terminate the Department of Geography at the end of the 1981–82 academic year.

From their experience of this decision, Kish and Ward offer comments on how departments of geography may counter similar attacks. They emphasize the teaching role, seeking to attract students in a competitive market and presenting 'our wares in a stimulating and excellent way'. This should involve examining

> the virtues of applied geography. As students become more conscious of careers, there may be a corresponding need to increase a skill-oriented curriculum. This could demonstrate the relevancy of geography and enhance its appeal to students (p. 14).

Kish and Ward – 'To appease some traditionalists' – claim that they do not advocate reorganizing the entire corpus of geography, 'but only attempt to broaden the appeal of geography to the student population'. (See also Ford, 1982; Powell, 1981.)

In research as well as teaching the trend towards geography as an applied discipline is being heavily emphasized, particularly in the United States, but not only there. Thus in announcing a new editorial policy for *The Professional Geographer*, the April 1982 issue of the AAG *Newsletter* reports that (p. 1)

> Brunn hopes to emphasize work in applied geography and to include information on corporation activities, state and local government projects, Federal government activities, activities by United Nations and other international agencies, and research projects.

The need for more applied work is widely accepted among American human geographers as a necessary means for the reproduction and even growth of their discipline. It takes a variety of forms (some argue, for example, for a greater concentration on cartography), as Mikesell (1980, p. 14) indicates:

> geographers fortunate enough to have secure teaching positions will worry about what they should do. The best response to this concern could be a decision to do what they have been doing, but with a keener appreciation of context and a greater willingness to be influenced by environment.

In Britain, the external pressures have not been quite as great. Nevertheless, there has been substantial expression of the need for applied geography, as a contribution from the discipline to the solution of current economic and social problems (see Chapter 6). Thus, for example, Bennett (1982, p. 69), writing of the 1982 conference of the Institute of British Geographers, found it

> possible to discern a strong and growing set of foci which, if they do not yet demark a new core, at least show an emerging commonality of interest. For this writer these foci were a widespread assertion of 'relevant' research, the reassertion of quantitative and analytical methods, and the rejection of recent anti-empirical movements.

Bennett welcomed these, noting that

hot [i.e. relevant] issues are not ones which British geographers are particularly noted for tackling *en masse* — and, as a result, the discipline has suffered a lack of public exposure, and a marked inability to influence public and private decisions.

and

At a time when higher education as a whole is under considerable challenge, and when geography as a discipline may suffer particularly severe pressure in some institutions it is heartening to see the emergence of concern with the hot issues expressed at this meeting (p. 71)

A generational model

The importance of context as an influence on the nature of human geography as a discipline has suggested that the paradigm model might be rephrased as a generational model (Johnston, 1978c, 1979b). This argues that changes in the external environment provide a necessary, but not sufficient, stimulus to changes within the discipline that can be interpreted as attempts to develop a new research programme, if not to launch a revolution and create a new normal science. Associated with the external changes must be a set of conditions within the discipline itself which is sympathetic to the new demands of the milieux. In most cases, these conditions are best met by younger members of the discipline. Stegmuller (1977, p. 148) agues that:

it is mostly young people who bring new paradigms into the world. And it is young people who are most inclined to champion new causes with religious fervour, to thump the propaganda drums.

In order to obtain influence, however, especially in times of resource shortage within higher education, such younger workers need the patronage of some established members of the discipline (see Chapter 1). According to Lemaine *et al.* (1976, p. 5):

Mendel's work, and that of his successors, was a response to scientific problems. But the scientific implications of their results were not pursued until there existed a strong group of scientists who, owing to their academic background and their position in the research community, were willing to abandon established conceptions.

Such reactions to environmental shifts may involve attempts to create a new paradigm or new research programme, or they may only require new branches of an existing paradigm/programme. Whichever it is, even when establishment support has been obtained, success is more likely to occur — or a bridgehead is more likely to be built, when certain criteria are met. These include (van den Daele and Weingart, 1976): an autonomous system of evaluation and reputation; an autonomous communication system; acknowledgement of the new ability to solve puzzles within the confines of the disciplinary matrix; a formal organization providing training programmes which allow reproduction and expansion of the new group's membership; an informal structure with leaders; and resources for research. In periods of academic expansion, these criteria rarely create problems; as Capel, Mikesell and others indicate, innovation is encouraged if it brings status, charisma and resources to the discipline. In periods of stagnation and retrenchment, the conditions are less favourable, and major shifts can only be achieved by revolutions among the existing members of the discipline. The generational model suggests that the latter is rare. With regard to the research record, Law (1976, p. 228) has counselled that:

> it may well be the case that scientists do lay special emphasis on the accounts in scientific papers, but my hunch is that there is immense (and non-trivial) varia-tion between scientists on this count. For some, science is something you do in the laboratory, something you talk about, and something you get excited about. For others, science is what they write and what they read in the journals. I would even hypothesize (in conformity with the invisible college notion) that those who are generally felt to be of higher status locate science less in their journals ·han in their own and other people's heads.

Many academics publish rarely, if ever, in their discipline's research journals; this is particularly so with academics over the age of ahout 40 (Chapter 1). For the latter, as changes occur in their discipline so they get 'left-behind'. The normal science that they continue to teach is based on the disciplinary matrix and exemplars into which they were socialized. But their colleagues, socialized later and influenced by subsequent environmental conditions, operate in different ways. The academic career cycle, in combination with a changing milieu, can produce a multi-paradigm teaching, if not research, discipline.

The key elements in the generational model (modified after Johnston, 1978c) are:

1   The external environment is a major influence on the contents of a disci-pline, especially one in the social sciences which is closely linked with that environment.

2   At times, the nature of this environment may shift. This provokes a reaction among a minority of members of the discipline who seek to stimu-late change in disciplinary practice by its established members and to generate interest in that change among the youngest generation of research workers. The latter is usually much more successful than the former.

3   Together, this grouping presents a new 'school of thought'. In some cases, opposing new schools may be stimulated.

4   The new school is coopted into the disciplinary career structure.

5   The publications of the new school come to dominate the disciplinary research output, as the productivity of the earlier generation declines.

6   Students face two or more separate generational schools in a depart-ment's teaching syllabus.

7   Over time, members of the new school attain seniority within the discipline.

Several consequences may follow from operation of this model. One is that if a discipline fails to react to a changing milieu, it will stagnate. Thus innovators are to be encouraged. Another is that some potential innovators may find that they are unable to influence their discipline, because the environment is not right and the establishment need not react to their suggestions. Duncan (1974, p. 109) has illustrated these 'processes of resistance to new scientific ideas within science itself', using the example of Hägerstrand's (1968) work on spatial diffusion. Hägerstrand's ideas were originally published in 1953, but Duncan's study of citations shows that widespread recognition was much delayed, compared to the average for all publications in geography (Stoddart, 1967a). The reasons for this, Duncan claims, were neither language nor Hägerstrand's relative isolation in Sweden, but the apparent irrelevance of his work to those steeped in another paradigm. Only when a new paradigm (spatial science) had been established was Hägerstrand's seminal contribution recognized:

> Hägerstrand's own attempts to disseminate his work met rejection from adherents to orthodoxy, but enthusiasm from the new geography group. The eventual relay of information to this community owed more to dogged personal effort than to the formal communication system of normal science, and general recognition was not achieved until professional allegiances were extremely reorganized (p. 130)

Even when a new idea has been recognized, however, those working on it may be divided into two or more virtually independent groups (Gatrell, 1982).

Human Geography: Paradigms, or Research Programmes, or . . .?

The previous sections have discussed various models of scientific and disciplinary development in the light of the recent situation in human geography and in the context of the detailed outline of intra-disciplinary debate, prsented in the preceding five chapters. The terminology of both Kuhn and Lakatos has been used. The present section attempts to bring the models and experience together.

The 1950s and 1960s
At the beginning of the period being studied here, regionalism (with its associated exceptionalist philosophy) dominated the discipline. There were several distinct branches within this, however, including within the United States the areal differentiation branch centred in the Midwest (mainly at the universities of Chicago and Wisconsin-Madison) and Sauer's Berkeley school; there was also a sizeable school based on Clark University, whose approach was close to that of the Midwest group (Bushong 1981; Prunty, 1979). In Britain, too, the regional theme dominated – though with a

stronger physical emphasis, and less concentration on particular 'schools' (Freeman, 1979). During the first decade or so after the war, the main developments were in systematic studies. Their stated (though often unrealized) aim was to increase the content of the regional descriptions, however, and not to break away from them. Understanding of particular places was to be advanced by increasing knowledge of the general processes operating to produce the unique characteristics.

This situation can be (and has been) interpreted as evidence for the paradigm model. The relevance of the generational version is indicated by the continued existence, into the 1960s, of work in the environmental determinism 'paradigm'. But regionalism is probably best described as a research programme. The negative heuristic, or hard core, was the desire to describe and provide explanatory accounts for identifiably separate components of the landscape (with British geographers paying more attention to the physical elements of the landscape than did their American counterparts). In the protective belt, there was a number of positive heuristics, or modes of investigation. With the development of systematic studies, these multiplied in number and their contacts with the hard core diminished. But the core remained, as a firm belief in – if not practice of – regional synthesis.

The 1950s and 1960s saw what Entrikin (1981, p. 1) describes as a 'transition between reigning orthodoxies . . . in which the spatial theme superseded the regional theme'. As Johnston and Gregory (1983) have outlined for the British case, and as was equally true in the American, this transition was marked by growth of separate systematic disciplines and a rapid distancing from the goal of regional synthesis by the new generation of both human and physical geographers. As Guelke (1977a, 1977b, 1978), Entrikin (1981) and others have indicated, initially the distancing was not rapid, and the main argument of several of the 'new geographers' was that quantitative analysis and the development of spatial science would provide both better techniques for identifying and describing regions and better understanding of the systematic elements of those regions.

To a considerable extent, therefore, it could be claimed that the developments of the 1950s chronicled in Chapter 3 of this book represented rapid extensions of the positive heuristic of the accepted core of the discipline. To this extent, therefore, a methodological shift has taken place (Chisholm, 1975) without a philosophical shift: it was what has been described elsewhere as a 'quiet revolution' (Johnston, 1978c, 1979b, 1981b). Increasingly, however, this shift was insufficient. Its link to the unfashionable regional approach was unattractive (Gould, 1979); the increased contact with other social sciences introduced systematic geographers to the excitements of the various topical specialisms; the generally high regard for the positivist social sciences; and the apparent need for a positivist spatial science: all these served to sever the links with the regional core and to encourage the establishment of human geography as a spatial science, as a searcher for order in the spatial organization of society with an emphasis on general laws and theories rather than on their relevance to particular places. The term region, when used, merely referred to a territorial division (as in regional science), not to a unique phenomenon-complex with its own personality.

The establishment and growth of spatial science has been identified by several commentators as the closest that human geography has come to being dominated by a single paradigm in the last few decades. But, as the discussion of Chapter 3 and the generational model outlined above indicates, there was never consensus. Spatial science is best interpreted as a major research programme, with its positivist core. With it are associated many branches, most of them focusing on particular systematic topics and a few being concerned mainly with methodological issues. And there can be little doubt that spatial science has been a progressive research programme – to some it still is. This is particularly illustrated by the technical achievements, with, for example, the move from simple linear models through multivariate linear and non-linear models, the adoption of systems analysis and the attack on the spatial autocorrelation problem to work on catastrophe and bifurcation theory. The texts illustrate this progression. (Compare, for example, Haggett, 1965c, with Haggett, Cliff and Frey, 1977 for a general view; Cliff and Ord, 1973 and Cliff and Ord, 1981, for a single methodological issue; and Hägerstrand, 1968, with Cliff *et al.*, 1981, for a particular topic.) And reviews suggest success. Haggett (1978, p. 161), writing on 'The spatial economy', claims that it

> is more carefully defined than before, we know a little more about its organization, the ways it responds to shocks, and the way some regional sections are tied to others. There now exist theoretical bridges, albeit incomplete and shaky, which span from pure, spaceless economics through to a more spatially disaggregated reality.

He, and many others, believe that this increased understanding, achieved through rigorous quantitative analysis, can make important contributions to public policy as an applicable quantitative geography (Bennett, 1981a).

Within this research programme there has been much launching of new branches, much shifting of individuals among those branches – often as the consequence of external environmental contacts (Pred, 1979; Johnston, 1982b). There have also been two major shifts, both contained within the research programme but generating considerable redirection within it. The first was the advocacy for a move from normative to behavioural approaches, as described in the first part of Chapter 5. As Cox (1981) indicates, this was not a departure from the hard core of the spatial science research programme: it was a change in emphasis and style. The other was the advocacy for 'welfare geography' described in Chapter 6. This, as Eyles and Smith (1978) describe it, was a response to environmental changes and a desire to make human geography more relevant to contemporary social problems. The shift was in topical emphasis, and the hard core of the programme remained.

The 1970s: increasing conflict

During the 1950s, 1960s and 1970s, therefore, a large research programme with many branches dominated the discipline of human geography. In the last of these decades, reaction to that research programme developed. One of these, the structuralist reaction outlined in Chapter 6, was a response to the changed external environment of the 1970s. So were the 'welfare geography' discussed in the previous paragraph and the call for 'quantitative geography and public policy' outlined above. Clearly, they were different reactions. The last two represented what are widely known as 'liberal reform' solutions. Problems have arisen and we must use our ability to describe those problems in order to solve them. In this sense, applied geography supported the current economic system, and sought to provide succour where problems arose (Johnston, 1981a). But the other reaction – the structuralist – perceived the problems as in the system itself and not just its current manifestations. It, too, sought to be relevant, but relevant to a different constituency. Instead of providing policy options for ameliorating problems, but not involving change in the processes that create such problems, it sought – by logical analysis – to show that only a change of economic system would lead to the permanent removal of the perceived problems.

This structuralist perspective could be presented as an alternative research programme, competing with spatial science. But Lakatos presents research programmes as scientific, in the positivist sense. Progressive research programmes are those which continue to enrich their empirical content through successful predictions, and researchers switch to them when they are aware of their superior predictive ability. The structuralist theories are not predictive in a positivist sense, and are neither verifiable nor falsifiable in the usually accepted context. They present not only a different disciplinary matrix but also, and even more so, a different world-view, a different perspective on social science. A switch to them, therefore, is an ideological-cum-philosophical decision, not a scientific one.

The emergence of the structuralist perspective as a reaction to both the environmental context and the support for that given by the content of human geography suggests that neither Kuhn's nor Lakatos's model is correct for human geography. (The relative failure of the structuralist perspective to gain a large foothold within human geography undoubtedly reflects the nature of its reaction, and the response of both the environmental milieu and its supporters within human geography to structuralism.) The concurrent emergence of the humanistic perspective backs this suggestion. This was very largely a development within human geography, although gaining much support from similar trends elsewhere in the social sciences. It was a reaction to the treatment of man in spatial science (including behavioural geography: Ley, 1981). It was also a return to, and reinterpretation of, earlier perspectives within human geography. This could be interpreted as the revivification of the dormant research programme. But again, the difference between the programmes is a difference of world-view, of philosophy not methodology. And, as with the structuralist approach, the humanist approach cannot be judged in terms of its predictive ability.

The situation in Anglo-American human geography in the early 1980s comprises a competition between three philosophies. It is not a competition between research programmes, on which judgements can be made using scientific criteria. As Claval (1981) notes, human geographers are currently debating epistemological issues, as well as technical and methodological ones. According to Ley (1981, p. 209):

> it is perhaps an indication of human geography's growing maturity as a social science that after its long period of empirical tranquility and reluctance to engage methodological, let along philosophical, issues, the subject has attempted over the past fifteen years to assimilate quantification, behavioralism and, more recently, humanism and structural marxism. This condensation of intellectual history, which in other disciplines was spread over a period of two or three times as long, has created a turbulence which, while exciting, can also be confusing.

This turbulence is indicated by the debates reviewed in the central chapters of this book.

To some extent, the three approaches to human geography – positivist, humanistic amd structuralist (Johnston, 1983b) – are paradigms, using that term to mean a 'world view', a philosophy with its own definition of knowledge and how it is obtained (i.e. its epistemology). And to some extent, individuals who switch from one to the other are involved in a personal revolution, making a major leap of faith from one definition of knowledge to another. But those leaps of faith are not scientifically based, in the positivist sense, and the humanistic and structuralist approaches cannot be judged on the same criteria as the positivist. In the social sciences, positivism is not necessarily the only source of knowledge. To many, understanding is more important than explanation – which latter may be impossible (Eyles and Lee, 1982).

Within each of these three approaches, Mulkay's ideas regarding intra-paradigm branching are clearly relevant (see p. 27). Within the humanistic approach, for example, the relative merits of phenomenology, idealism and existentialism are being explored, and the great range of interests and methodologies within the positivist approach has already been outlined. This exploration involves human geographers working with the literature, and increasingly the members, of the other social sciences. New (to human geographers) ideas are identified, and attempts are made to transfer them into geographical contexts. Some are rapidly discarded – leading Ley (1981, p. 209) to claim that 'In true North American fashion, obsolescence is setting in more and more speedily'. Some are adopted, but with the geographical importers often apparently unaware of controversies regarding those ideas within the source areas and undertaking less than complete evaluations of their new ideas (Agnew and Duncan, 1981; Duncan, 1980).

This 'foreign trade' of ideas stimulates the intra-disciplinary turbulence. (Note that some, such as Eliot Hurst, 1980, seek to dedefine disciplines such as human geography and to integrate them around the single social science of historical materialism. Over time, this should lead to a different mode of of socialization that would preclude such transfers of ideas . . . if it were allowed to happen by the established strength of the current disciplinary establishments.) And part of that turbulence involves the search for reconciliations between philosophies (see Harrison and Livingstone, 1982; Hay, 1979a; Johnston, 1980, 1982c; Livingstone and Harrison, 1981). There would appear to be more scope for combining aspects of the structur-: alist and humanistic approaches (Eyles, 1981; Johnston, 1983b, 1983c), than for combining either of those with positivism. (Which does not exclude the use of quantitative methods in any of the three. These are not exclusive to positivism, although they are generally associated with that approach.) The structuralist approach investigates the basic processes within society. The humanistic approaches study the actions of individuals. A combination of the two sets individuals within a context which is both constraining and enabling (see Johnston, 1983b, 1983c), whereas phenomenological study of geography can investigate why we study what we do.

The discourses of human geography

It is not surprising that Kuhn's paradigm model has become so popular among human geographers and other social scientists, for it offers a framework within which certain disciplinary changes can be accommodated (if not fitted) and it provides a propaganda base for the exhortation of further changes. In a very similar way, the concept of the research programme introduced by Lakatos could be as attractive, although it has not proven so: Kuhn, perhaps because he was first in the field, has received much more publicity. But neither is sufficient for an outline of the history of human geography for the simple reason, as both of their originators fully realized and stated, that they are tied to one epistemology only. The world-views of paradigms and of research programmes are those of positivism, as are their evaluation procedures. Non-positivist approaches to knowledge cannot be evaluated on positivist criteria. Hence in a discipline, or part of discipline, that contains such non-positivist approaches, the models of Kuhn and Lakatos are inappropriate.

In a science, progress is linear, so that knowledge in time becomes obsolete. This has been shown in many scientific disciplines by studies of citation patterns, including the length of time over which a piece of work is cited. Once it is no longer cited, it is obsolete, having been replaced (and expanded on) by more recent work. The lapse rates in most sciences are generally short, with an average of only years before papers cease to be cited. This tends to be the case in geography too (Stoddart, 1967a).

But it need not be so, and indeed some ideas in human geography are never made obsolete, even though they may fall into disfavour for a while. Popper

(1967) has introduced the concept of three separate, though interdependent, worlds:

1   World One – the physical world of objective knowledge (which corresponds with Kirk's concept of the phenomenal environment – p. 141).

2   World Two – the subjective world of conscious experiences (the behavioural environment).

3   World Three – the knowledge stored in books, computers and like objects, that themselves are in World One.

As Brid (1975) points out, the publications of geographical and other writers are lodged in World Three, and are available for future generations of researchers, whether or not teachers direct them towards such material during the socialization process. Literature surveys may unearth ideas that are thought worth reviving, as with the anarchist notions of a nineteenth-century geographer, Peter Kropotkin (Stoddart, 1975b; Breitbart, 1981). Branches of the discipline, then, may never actually die. They merely fall dormant, to await possible revival as the result of either a chance discovery or a detailed tracing of academic roots in World Three.

In the social sciences, three basic sets of philosophies have dominated: these are the positivist (with which empiricism is associated), humanistic and structuralist. All three are now vigorously pursued and debated within human geography. It is unlikely that one will ever achieve dominance (unless there is some form of totalitarian institutionalization of one only). At particular times, one may be more important than at others, as the result of interactions between scholars and their environments. But exploration of World Three is not constrained in the social sciences, and researchers are not as strictly socialized into a particular philosophy as they are in the natural sciences. Thus (Johnston, 1981b, pp. 313–14):

> Much geographical work is exploratory, and is conducted by individuals who operate independently. Indeed, many, although influenced by what they read, are in no sense socialized into a particular matrix or set of exemplars which might be associated with a 'research school' and its leader. The charisma of certain individuals and their published works may occasionally produce the prophet and disciple situation. Much more usual, however, is a situation of fickle allegiances.

Such fickle allegiances usually involve movement among the various branches of a particular philosophy (see Pred, 1979; Johnston, 1982b), but may involve experimentation with different philosophies. The latter can involve 'moving back' through World Three, seeking to revive something that is currently unfashionable. (Note recent attempts to resurrect regional geography as the central pillar of the discipline's educational contribution: cf Steel, 1982; Hart, 1982.) Such wanderings suggest anarchy, in the sense used by Feyerabend (p. 26: Johnston, 1976b, 1978c); this is not chaos, but a matrix of shifting small communities, independent of central control.

How can such a shifting matrix best be described? Of the various approaches outlined in Chapter 1, that proposed by Foucault appears to be the most promising, because it is not tied to a positivist definition of scientific progress and suggests no precise criteria for identifying shifts in disciplinary content. (Indeed, Foucault's concepts of discourse and *episteme* pay scant regard to the institutionalized academic disciplines.) As indicated in Chapter 1, Foucault gives no clues as to how the *episteme* or world-view of a society changes, nor as to how discourses wax and wane. But the concept of the discourse and its relationship to an *episteme* appears more valid for a social science such as human geography that do the disciplinary matrix plus exemplars concept of Kuhn and the research programmes idea of Lakatos, both of which are set in an unchanging *episteme* – that of positivism.

Only Claval (1981) of geographers interested in such issues makes detailed reference to Foucault. (Scott's, 1982, work shows similar tendencies.) He notes that the history of geographical thought can be presented in one of two ways. In the first, 'it attempts to recreate the past' (p. 230) and to link geography to the society in which it flourished. In the second, it 'attempts to work out the logical development of ideas' (p. 230), showing how various disciplines are linked together:

> It . . . questions knowledge by going beyond the statement of facts [i.e. description of the discourse]: it wonders what authority was behind them and penetrates the development of the discipline from the inside (p. 230).

Too many historians of geography, he claims, are 'prisoners of a limited field: they ignore the evolution of thought in parallel disciplines' (p. 231). A full history must set the study of discourse in the context of the *episteme*.

Claval interprets Foucault's work as suggesting that different periods of (European) history have been characterized by particular world-views. The concept of knowledge has changed several times. At the present time, it seems, a transition is taking place, from the classical *episteme*, with its positivist base and its concentration on appearances, to the modern episteme

> characterized by the search, through the exploration of the underlying layers of reality, for causalities which express the fundamental play of forces (p. 236).

This shift began in the early nineteenth century; it reached Anglo-American human geography in the late 1960s!

Foucault's work and Claval's interpretations are suggestive rather than definitive with regard to the recent history of human geography. The concept of a discourse provides a useful tool for describing the ties of a community of scholars, who may cross the institutionalized disciplinary boundaries – as illustrated in many areas of contemporary human geography which have strong links with some members of several other disciplines. Both Kuhn's paradigm as disciplinary matrix and Lakatos' research programme could be interpreted in this way, however, as, at a smaller scale, could the paradigm as exemplar concept, for several communities follow an exemplar from another discipline. In this descriptive sense, the benefit of the concept of discourse is simply that it is non-positivist.

Critics of Kuhn and Lakatos have pointed out that both present academic disciplines as isolated communities with their own, scientific, world view. Foucault, on the other hand, sets those communities in the contexts of the societal world-view, or *episteme*. In recent years, a number of human geographers have indicated the need to study the discipline-environment interaction more closely. Foucault's framework of interactions between discourse and *episteme* suggests a model for such a task. According to this, however, there is a dominant – if not predominant – *episteme* at any one period. Claval identifies these as: sixteenth-century idealism; seventeenth and eighteenth-ceuntury classical positivism; and nineteenth and twentieth century modern realism/structuralism. If this is so, then how does one account for the recent revival of idealism and other humanistic epistemologies in human geography, the late introduction and continued strength of positivism, and the relatively small success of structuralism in the 1970s?

There can be little doubt that human geography has been relatively isolated from the mainstream of philosophical discussion in the social sciences in recent years. This in itself may account for the current exploration of World Three and the lack of consensus about a philosophy for the discipline. Even if the disciplinary framework is abandoned, however, a wider evaluation of contemporary social sciences will show a total lack of epistemological consensus. Several *epistemes* exist within the social sciences. They are competing for followers, and in part this is being done by the establishment of discourses in the various disciplines – with some of the discourses having the avowed intention of removing the disciplines.

It may be that the idea of a single *episteme* – a spirit of the times – is relevant for periods prior to that being studied here. It is doubtful, however; certainly so for the last century. The present situation is one of several *epistemes*, with links to discourses in various social science disciplines. Currently, human geography has links to at least three of these. One of them – the positivist – has a world-view that is closely allied with the world-view (or ideology) of Anglo-American capitalist society. The other two reflect world-views which are very much in conflict with that ideology. It may be that positivist spatial science will endure, and maintain its strength within human geography, for as long as the present societal system. (Note that this suggests that it is the classical *episteme* that is dominant, not the modern one that has succeeded it according to Foucault and Claval.)

At the present time, then, human geography comprises three discourses each linked to a major world-view, or *episteme*. Within the discipline, each of these operates its own set of rules. The structure of each is very much like that of a research programme. There is a hard core which defines the discourse in terms of the *episteme* and which controls its work. Surrounding it is a protective belt of small communities of workers – among which individual researchers move – that carries the discourse forwards, through continued research activity. These various discourses are linked to their *epistemes* and they seek to expand, by attracting more workers who accept their world view. Individuals are invited to undergo personal revolutions.

And the future?

Human geography has always been somewhat pluralistic as a discipline, containing more than one discourse. It is undoubtedly pluralistic at the present time. No one discourse controls the institutional organization of the discipline – its learned societies and its pedagogic bodies. And none is likely to in the foreseeable future. (This assumes of course, that power within society allows pluralisism to continue. In Chile in the late 1970s/early 1980s all social science disciplines that did not conform to the government's world-view were abolished in the universities, as were those unacceptable discourses within the disciplines that were retained.)

The pluralism that currently characterizes human geography is to some a polarization: only one *episteme* can be correct. To others this is not so, and some accommodation between *epistemes* should be engineered through the intersection of discourses. While this situation remains, debate will continue, not only with regard to research but also teaching strategies (Gould, 1977, 1978, 1981). And, of course, debate will continue within the individual discourses. In the early 1970s, a sample of the discipline's past iconoclasts presented their views of the future human geography (Chorley, 1973). Much of what has happened in the last decade was not foreshadowed there. This may be a criticism of the sampling; it may reflect the unpredict-ability of social science; it may indicate that the environmental changes of the last decade were unforeseen. There can be little doubt of the relevance of the last of these for the future.

In the first edition of this book, and elsewhere, human geography was characterized as 'branching towards anarchy'. As indicated earlier in this edition, anarchy implies

> free association and voluntary cooperation of individuals and groups (Labedz, 1977, p. 22)

which is not a synonym of chaos. This trend continues. Human geography as a discipline within the institutionalized discipline of geography remains, and will continue to do so. Within it, there are now three major discourses, each with several subsidiary discourses: in addition, there are attempts to create new discourses by linking the three. There is no normal science, no consensus over a disciplinary matrix, no agreement over the right exemplars. Human geography is a shifting conglomerate of small communi-ties, with which many individuals are only weakly linked. No change in this situation seems likely.

Bibliography

ABLER, R. F. 1971: Distance, intercommunications, and geography. *Proceedings, Association of American Geographers* **3**, 1–5.

ABLER, R. F., ADAMS, J. S. and GOULD, P. R. 1971: *Spatial organization: the geographer's view of the world.* Englewood Cliffs, NJ: Prentice-Hall.

ACKERMAN, E. A. 1945: Geographic training, wartime research, and immediate professional objectives. *Annals, Association of American Geographers* **35**, 121–43.

—— 1958: *Geography as a fundamental research discipline.* University of Chicago, Department of Geography Research Paper 53.

—— 1963: Where is a research frontier? *Annals, Association of American Geographers* **53**, 429–40.

ADAMS, J. S. 1969: Directional bias in intra-urban migration. *Economic Geography* **45**, 302–23.

AGNEW, J. A. and DUNCAN, J. S. 1981: The transfer of ideas into Anglo-American human geography *Progress in Human Geography* **5**, 42–57.

ALEXANDER, D. 1979: Catastrophic misconception? *Area* **11**, 228–30.

ALEXANDER, J. W. and ZAHORCHAK, G. A. 1943: Population-density maps of the United States: techniques and patterns. *Geographical Review* **33**, 457–66.

AMEDEO, D. and GOLLEDGE, R. G. 1975: *An introduction to scientific reasoning in geography.* New York: John Wiley.

ANDERSON, J. 1973: Ideology in geography: an introduction. *Antipode* **5**(3), 1–6.

APPLETON, J. 1975: *The experience of landscape.* London: John Wiley.

ARCHER, J. C. and TAYLOR. P. J. 1981: *Section and party.* Chichester: John Wiley.

BADCOCK, B. A. 1970: Central-place evolution and network development in south Auckland, 1840–1968: a systems analytic approach. *New Zealand Geographer* **26**, 109–35.

BAKER, A. R. H. 1972: Rethinking historical geography. In A. R. H. Baker (ed.), *Progress in Historical Geography* (Newton Abbott: David & Charles), 11–28.

BALLABON, M. B. 1957: Putting the 'economic' into economic geography. *Economic Geography* **33**, 217–23.

BARNES, B. 1974: *Scientific knowledge and sociological theory.* London: Routledge & Kegan Paul.

—— 1982: *T. S. Kuhn and social science.* London: Macmillan.

BARROWS, H. H. 1923: Geography as human ecology. *Annals, Association of American Geographers* **13**, 1–14.

BATTY, M. 1976: *Urban modelling: algorithms, calibrations, predictions.* London: Cambridge University Press.

— 1978: Urban models in the planning process. In D. T. Herbert and R. J. Johnston (eds.), *Geography and the urban environment,* vol. I (London: John Wiley), 63–134.

BELL, D. 1973: *The coming of post-industrial society.* New York: Basic Books.

BEAUMONT, J. R. and GATRELL, A. C. 1982: *An introduction to Q-Analysis.* CATMOG 34, Geo Abstracts Ltd, Norwich.

BENNETT, R. J. 1974: Process identification for time-series modelling in urban and regional planning. *Regional Studies* **8**, 157–74.

— 1975: Dynamic systems modelling of the Northwest region: 1. Spatio-temporal representation and identification. 2. Estimation of the spatio-temporal policy model. 3. Adaptive parameter policy model. 4. Adaptive spatio-temporal forecasts. *Environment and Planning A* **7**, 525–38, 539–66, 617–36, 887–98.

— 1978a: Forecasting in urban and regional planning closed loops: the examples of road and air traffic forecasts. *Environment and Planning A* **10**, 145–62.

— 1978b: *Spatial time series: analysis, forecasting and control,* London: Pion.

— 1979: Space-time models and urban geographical research in D. T. Herbert and R. J. Johnston (eds.), *Geography and the urban environment: progress in research and application,* vol. 2, (London: John Wiley), 27–58.

— 1981a: Quantitative geography and public policy. In N. Wrigley and R. J. Bennett (eds.), *Quantitative geography* (London: Routledge & Kegan Paul), 387–96.

— 1981b: A hierarchical control solution to allocation of the British Rate Support Grant. *Geographical Analysis* **13**, 300–14.

— (ed.) 1981c: *European progress in spatial analysis.* London: Pion.

— 1981d: Quantitative and theoretical geography in Western Europe. In R. J. Bennett (ed.), *European progress in spatial analysis* (London: Pion), 1–32.

— 1982: Geography, relevance and the role of the Institute. *Area* **14**, 69–71.

BENNETT, R. J. and CHORLEY, R. J. 1978: *Environmental systems: philosophy, analysis and control.* London: Methuen.

BENNETT, R. J. and WRIGLEY, N. 1981: Introduction. In N. Wrigley and R. J. Bennett (eds.), *Quantitative geography* (London: Routledge & Kegan Paul), 3–11.

BERDOULAY, V. 1981: The contextual approach. In D. R. Stoddart (ed.), *Geography, ideology and social concern* (Oxford: Blackwell,) 8–16.

BERRY, B. J. L. 1958: A critique of contemporary planning for business centers. *Land Economics* **25**, 306–12.

— 1959a: Ribbon developments in the urban business pattern. *Annals, Association of American Geographers* **49**, 145–55.

— 1959b: Further comments concerning 'geographic' and 'economic' economic geography. *The Professional Geographer* **11**(1), 11–12.

— 1964a: Cities as systems within systems of cities. *Papers, Regional Science Association* **13**, 147–63.

— 1964b: Approaches to regional analysis: a synthesis. *Annals, Association of American Geographers* **54**, 2–11.

—— 1965: Research frontiers in urban geography. In P. M. Hauser and L. F. Schnore (eds.), *The study of urbanization* (New York: John Wiley), 403–30.

—— 1966: *Essays on commodity flows and the spatial structure of the Indian economy.* University of Chicago, Department of Geography, Research Paper 111.

—— 1967: *The geography of market centers and retail distribution.* Englewood Cliffs, NJ: Prentice-Hall.

—— 1968: A synthesis of formal and functional regions using a general field theory of spatial behavior. In B. J. L. Berry and D. F. Marble (eds.), *Spatial analysis* (Englewood Cliffs, NJ: Prentice-Hall), 419–28.

—— (ed.) 1971: *Comparative factorial ecology. Economic Geography* **47**, 209–367

—— 1972a: Hierarchical diffusion: the basis of development filtering and spread in a system of growth centers. In N. M. Hansen (ed.), *Growth centers in regional economic development* (New York: The Free Press), 108–38.

—— 1972b: 'Revolutionary and counter-revolutionary theory in geography' – a ghetto commentary, *Antipode* **4**(2), 31–3.

—— 1972c: More on relevance and policy analysis, *Area* **4**, 77–80.

—— 1973a: *The human consequences of urbanization.* London: Macmillan.

—— 1973b: A paradigm for modern geography. In R. J. Chorley (ed.), *Directions in Geography* (London: Methuen), 3–22.

—— 1974a: Review of H. M. Rose (ed.), *Perspectives in geography 2. Geography of the ghetto, perceptions, problems and alternatives. Annals, Association of American Geographers* **64**, 342–5.

—— 1974b: Review of David Harvey, *Social Justice and the City. Antipode* **6**(2), 142–5, 448.

—— 1978a: Introduction: a Kuhnian perspective. In B. J. L. Berry (ed.), *The nature of change in geographical ideas* (de Kalb: Northern Illinois University Press), vii-x.

—— 1978b: Geographical theories of social change. In B. J. L. Berry (ed.), *The nature of change in geographical ideas* (de Kalb: Northern Illinois University Press), 17–36.

BERRY, B. J. L. and BAKER, A. M. 1968: Geographic sampling. In B. J. L. Berry and D. F. Marble (eds.), *Spatial analysis* (Englewood Cliffs, NJ: Prentice-Hall), 91–100.

BERRY, B. J. L. and GARRISON, W. L. 1958c: The functional bases of the central place hierarchy. *Economic Geography* **34**, 145–54.

—— 1958b: Recent developments in central place theory. *Papers and Proceedings, Regional Science Association* **4**, 107–20.

BERRY, B. J. L. and HORTON, F. E. (eds.) 1974: *Urban environmental management: planning for pollution control.* Englewood Cliffs, NJ: Prentice-Hall.

BERRY, B. J. L. *et al.* 1974: *Land use, urban form and environmental quality.* Department of Geography, Research Paper 155, University of Chicago.

VON BERTALANFFY, L. 1950: An outline of general systems theory. *British Journal of the Philosophy of Science* **1**, 134–65.

BILLINGE, M. 1977: In search of negativism: phenomenology and historical geography. *Journal of Historical Geography* **3**, 55–68.

BILLINGE, M., GREGORY, D. and MARTIN, R. L. (eds.) 1982: *Reflections on a revolution*. London: Macmillan.

BIRD, J. H. 1975: Methodological implications for geography from the philosophy of K. R. Popper. *Scottish Geographical Magazine* **91**, 153–63.

—— 1977: Methodology and philosophy. *Progress in Human Geography* **1**, 104–10.

—— 1978: Methodology and philosophy. *Progress in Human Geography* **2**, 133–40.

BLAIKIE, P. M. 1978: The theory of the spatial diffusion of innovations: a spacious cul-de-sac. *Progress in Human Geography* **2**, 268–95.

BLAUG, M. 1975: Kuhn versus Lakatos, or paradigms versus research programmes in the history of economics. *History of Political Economy* **7**, 399–419.

BLAUT, J. M. 1979: The dissenting tradition. *Annals, Association of American Geographers* **69**, 157–64.

BLOWERS, A. T. 1972: Bleeding hearts and open values. *Area* **4**, 290–2.

—— 1974: Relevance, research and the political process *Area* **6**, 32–6.

BLUMENSTOCK, D. I. 1953: The reliability factor in the drawing of isarithms. *Annals, Association of American Geographers* **43**, 289–304.

BODDY, M. J. 1976: The structure of mortgage finance: building societies and the British social formation. *Transactions, Institute of British Geographers* **NS1**, 58–71.

BOYLE, M. J. and ROBINSON, M. E. 1979: Cognitive mapping and understanding. In D. T. Herbert and R. J. Johnston (eds.) *Geography and the urban environment*, volume 2 (Chichester: John Wiley), 59–82.

BREITBART, M. M. 1981: Peter Kropotkin, the anarchist geographer. In D. R. Stoddart (ed.), *Geography, ideology and social concern* (Oxford: Blackwell), 134–53.

BRITTAIN, S. 1977: Economic liberalism. In A. Bullock and O. Stallybrass (eds.), *The Fontana dictionary of modern thought* (London: Fontana Books), 188–9.

BROOKFIELD, H. C. 1962: Local study and comparative method: an example from Central New Guinea. *Annals, Association of American Geographers* **52**, 242–54.

—— 1964: Questions on the human frontiers of geography. *Economic Geography* **40**, 283–303.

—— 1969: On the environment as perceived. In C. Board *et al.* (eds.), *Progress in Geography* **1**, London: Edward Arnold, 51–80.

—— 1973: On one geography and a Third World. *Transactions, Institute of British Geographers* **58**, 1–20.

—— 1975: *Interdependent development*. London: Methuen.

BROWN, L. A. 1968: *Diffusion processes and location: a conceptual framework and bibliography*. Regional Science Research Institute, Bibliography Series 3, Philadelphia.

—— 1975: The market and infrastructure context of adoption: a spatial perspective on the diffusion of innovation. *Economic Geography* **51**, 185–216.

—— 1981: *Innovation diffusion: a new perspective*. London. Methuen.

BROWN, L. A. and MOORE, E. G. 1970: The intra-urban migration process: a perspective. *Geografiska Annaler* **52B**, 1–13.

BROWN, R. H. 1943: *Mirror for Americans: likeness of the eastern seaboard 1810.* New York: American Geographical Society.

BROWN, S. E. 1978: Guy-Harold Smith, 1895–1976. *Annals, Association of American Geographers* **68**, 115–18.

BROWNING, C. 1982: *Conversations with geographers: career pathways and research styles.* University of North Carolina at Chapel Hill, Department of Geography, Occasional Paper 16.

BRUNN, S. D. 1974: *Geography and politics in America.* New York: Harper & Row.

BRUSH, J. E. 1953: The hierarchy of central places in southwestern Wisconsin. *Geographical Review* **43**, 380–402.

BUCHANAN, K. 1962: West wind, east wind. *Geography* **47**, 333–46.

—— 1973: The white north and the population explosion. *Antipode* **5**(3), 7–15.

BULLOCK, A. 1977: Liberalism. In A. Bullock and O. Stallybrass (eds.), *The Fontana dictionary of modern thought* (London: Fontana Books), 347.

BUNGE, W. 1962: second edition, 1966. *Theoretical geography.* Lund Studies in Geography, Series C 1, Lund: C. W. K. Gleerup.

—— 1968: Fred K. Schaefer and the science of geography. *Harvard Papers in Theoretical Geography,* Special Papers Series, Paper A, Laboratory for Computer Graphics and Spatial Analysis, Harvard University, Cambridge, Mass.

—— 1971: *Fitzgerald: geography of a revolution.* Cambridge, Mass: Schlenkman.

—— 1973a: Spatial prediction. *Annals, Association of American Geographers* **63**, 566–8.

—— 1973b: Ethics and logic in geography. In R. J. Chorley (ed.), *Directions in geography* (London: Methuen), 317–31.

—— 1973c: The geography of human survival. *Annals, Association of American Geographers* **63**, 275–95.

—— 1973d: The geography. *The Professional Geographer* **25**, 331–7.

—— 1979: Fred K. Schaefer and the science of geography. *Annals, Association of American Geographers* **69**, 128–33.

BUNGE, W. and BORDESSA, R. 1975: *The Canadian alternative: survival, expeditions and urban change.* Geographical Monographs, Atkinson College, York University. Downsview, Ontario.

BUNTING, T. E. and GUELKE, L. 1979: Behavioral and perception geography: a critical appraisal. *Annals, Association of American Geographers* **69**, 448–62, 471–4.

BURNETT, K. P. (ed.) 1981: *Studies in choice, constraints, and human spatial behavior. Economic Geography* **57**, 291–383.

BURTON, I. 1963: The quantitative revolution and theoretical geography. *The Canadian Geographer* **7**, 151–62.

BURTON, I., KATES, R. W. and WHITE, G. F. 1978: *The environment as hazard.* New York: Oxford University Press.

BUSHONG, A. D. 1981: Geographers and their mentors: a genealogical view of American academic geography. In B. W. Blouet (ed.), *The origins of academic geography in the United States* (Hamden, Conn: Archon Books), 193–220.

BUTLIN, R. A. 1982: *The transformation of rural England c. 1580-1800.* Oxford: Oxford University Press.

BUTTIMER, A. 1971: *Society and milieu in the French geographical tradition.* Chicago: Rand McNally.

—— 1974: *Values in geography.* Commission on College Geography, Resource Paper 24, Association of American Geographers, Washington.

—— 1976: Grasping the dynamism of lifeworld. *Annals, Association of American Geographers* **66**, 277–92.

—— 1978a: Charism and context: the challenge of La Géographie Humaine. In D. Ley and M. S. Samuels (eds.), *Humanistic geography: prospects and problems.* Chicago: Maaroufa Press, 58–76.

—— 1978b: On people, paradigms and progress in geography. Institutionen for Kulturgeografi och Economisk Geografi vid Lunds Universitet, *Rapporter och Notiser* **47**.

—— 1979: Erewhon or nowhere land. In S. Gale and G. Olsson (eds.), *Philosophy in geography* (Dordrecht: D. Reidel), 9–38.

—— 1981: On people, paradigms and 'progress' in geography In D. R. Stoddart (ed.), *Geography, ideology and social concern* (Oxford: Blackwell), 70–80.

—— 1983: *The practice of geography.* London: Longman.

BUTTIMER, A. and HÄGERSTRAND, T. 1980: *Invitation to dialogue: a progress report.* DIA Paper 1, University of Lund, Lund.

CAMERON, I. 1980: *To the farthest ends of the earth.* London: Macdonald.

CAPEL, H. 1981: Institutionalization of geography and strategies of change. In D. R. Stoddart (ed.), *Geography, ideology and social concern* (Oxford: Blackwell), 37–69.

CAREY, H. C. 1858: *Principles of social science.* Philadelphia: J. Lippincott.

CARLSTEIN, T. 1980: *Time, resources, society and ecology.* Department of Geography, University of Lund, Lund.

CARLSTEIN, T., PARKES, D. N. and THRIFT, N. J. (eds.): *Timing space and spacing time* (three volumes). London: Edward Arnold.

CARROLL, G. R. 1982: National city-size distributions: what do we know after 67 years of research? *Progress in Human Geography* **6**, 1–43.

CARROTHERS, G. A. P. 1956: An historical review of the gravity and potential concepts of human interaction. *Journal, American Institute of Planners* **22**, 94–102.

CHAPMAN, G. P. 1977: *Human and environmental systems: a geographer's appraisal.* London: Academic Press.

CHAPPELL, J. E., Jr 1975: The ecological dimension: Russian and American views. *Annals, Association of American Geographers* **65**, 144–62.

—— 1976: Comment in reply. *Annals, Association of American Geographers* **66**, 169–73.

CHAPPELL, J. M. A. and WEBBER, M. J. 1970: Electrical analogues of spatial diffusion processes. *Regional Studies* **4**, 25–39.

CHISHOLM, M. 1962: *Rural settlement and land use.* London: Hutchinson.

—— 1966: *Geography and economics.* London: G. Bell & Sons.

—— 1967: General systems theory and geography. *Transactions, Institute of British Geographers* **42**, 45–52.

—— 1971a: In search of a basis for location theory: micro-economics or welfare economics? In C. Board *et al.* (eds.), *Progress in Geography* **3**, London: Edward Arnold, 111–34.

—— 1971b: Geography and the question of 'relevance'. *Area* **3**, 65–8.

—— 1973: The corridors of geography. *Area* **5**, 43.

—— 1975: *Human geography: evolution or revolution?* Harmondsworth: Penguin Books.

—— 1976: Regional policies in an era of slow population growth and higher unemployment. *Regional Studies* **10**, 201–13.

CHISHOLM, M., FREY, A. E. and HAGGETT, P. (eds.), 1971: *Regional forecasting*. London: Butterworth.

CHISHOLM, M. and MANNERS, G. (eds.) 1973: *Spatial policy problems of the British economy*. London: Cambridge University Press.

CHISHOLM, M. and O'SULLIVAN, P. 1973: *Freight flows and spatial aspects of the British economy*. Cambridge: CUP.

CHORLEY, R. J. 1962: Geomorphology and general systems theory. *Professional Paper* 500-B, United States Geological Survey, Washington.

—— 1964: Geography and analogue theory. *Annals, Association of American Geographers* **54**, 127–37.

—— 1973a: Geography as human ecology. In R. J. Chorley (ed.), *Directions in geography* (London: Methuen), 155–70.

—— (ed.) 1973b: *Directions in geography*. London: Methuen.

CHORLEY, R. J. and BENNETT, R. J. 1981: Optimization: control models. In N. Wrigley and R. J. Bennett (eds.), *Quantitative geography* (London: Routledge & Kegan Paul), 219–24.

CHORLEY, R. J. and HAGGETT, P. 1965a: Trend-surface mapping in geographical research. *Transactions and Papers, Institute of British Geographers* **37**, 47–67.

—— (eds.) 1965b: *Frontiers in geographical teaching*. London: Methuen.

—— (eds.) 1967: *Models in geography*. London: Methuen.

CHORLEY, R. J. and KENNEDY, B. A. 1971: *Physical geography: a systems approach*. London: Prentice-Hall International.

CHRISTALLER, W. 1966: *Central places in southern Germany* (translated by C. W. Baskin). Englewood Cliffs, NJ: Prentice-Hall.

CHRISTENSEN, K. 1982: Geography as a human science. In P. Gould and G. Olsson (eds.), *A search for common ground* (London; Pion), 37–57.

CLARK, A. H. 1954: Historical geography. In P. E. James and C. F. Jones (eds.), *American geography: inventory and prospects* (Syracuse: Syracuse University Press), 70–105.

—— 1977: The whole is greater than the sum of the parts: a humanistic element in human geography. In D. R. Deskins *et al.* (eds.), *Geographic humanism, analysis and social action: a half century of geography at Michigan* (Michigan Geographical Publication No. 17, Ann Arbor), 3–26.

CLARK, D., DAVIES, W. K. D. and JOHNSTON, R. J. 1974: The application of factor analysis in human geography. *The Statistician* **23**, 259–81.

CLARK, K. G. T. 1950: Certain underpinnings of our arguments in human geography. *Transactions, Institute of British Geograpers* **16**, 15–22.

CLARK, W. A. V. 1975: Locational stress and residential mobility in a New Zealand context. *New Zealand Geographer* **31**, 67–79.

—— 1981: Residential mobility and behavioral geography: parallelism or interdependence? In K. R. Cox and R. G. Golledge (eds.), *Behavioral problems in geography revisited* (London: Methuen), 182–205.

CLARKSON, J. D. 1970: Ecology and spatial analysis. *Annals, Association of American Geographers* **60**, 700–16.

CLAVAL, P. 1981: Epistemology and the history of geographical thought. In D. R. Stoddart (ed.), *Geography, ideology and social concern* (Oxford: Blackwell), 227–39.

CLIFF, A. D. and ORD, J. K. 1973: *Spatial autocorrelation.* London: Pion.

—— 1981: *Spatial process* London: Pion.

CLIFF, A. D. *et al.* 1975: *Elements of spatial structure: a quantitative approach.* London: Cambridge University Press.

CLIFF, A. D. *et al.* 1981: *Spatial diffusion,* Cambridge: Cambridge University Press.

COATES, B. E., JOHNSTON, R. J. and KNOX, P. L. 1977: *Geography and inequality.* Oxford: Oxford University Press.

COFFEY, W. J. 1981: *Geography: towards a general spatial systems approach.* London: Methuen.

COLE, J. P. 1969: Mathematics and geography. *Geography* **54**, 152–63.

COLE, J. P. and KING C. A. M. 1968: *Quantitative geography.* London: John Wiley.

CONZEN, M. P. 1981: The American urban system in the nineteenth century. In D. T. Herbert and R. J. Johnston (eds.), *Geography and the urban environment,* vol. 4 (Chichester: John Wiley), 295–348.

COOMBES, M.G. *et al.* 1982: Functional regions for the population census of Great Britain. In D. T. Herbert and R. J. Johnston (eds.), *Geography and the urban environment,* vol. 5 (Chichester: John Wiley), 63–111.

COOKE, R. U. and ROBSON, B. T. 1976: Geography in the United Kingdom, 1972–1976. *Geographical Journal* **142**, 3–22.

COOPER, W. 1952: *The struggles of Albert Woods.* London: Jonathan Cape.

COPPOCK, J. T. 1974: Geography and public policy: challenges, opportunities and implications. *Transactions, Institute of British Geographers* **63**, 1–16.

COURT, A. 1972: All statistical populations are estimated from samples. *The Professional Geographer* **24**, 160–1.

COX, K. R. 1969: The voting decision in a spatial context. In C. Board *et al.* (eds.), *Progress in Geography* **1** (London: Edward Arnold), 81–118.

—— 1973: *Conflict, power and politics in the city: a geographic view.* New York: McGraw Hill.

—— 1976: American geography: social science emergent. *Social Science Quarterly* **57**, 182–207.

—— 1979: *Location and public problems.* Oxford: Basil Blackwell.

—— 1981: Bourgeois thought and the behavioral geography debate. In K. R. Cox and R. G. Golledge (eds.), *Behavioral problems in geography revisited* (London: Methuen), 256–79.

COX, K. R. and GOLLEDGE, R. G. 1969: Editorial introduction: behavioral models in geography. In K. R. Cox, and R. G. Golledge (eds.), *Behavioral problems in geography: a symposium* (Northwestern University Studies in Geography 17, Evanston), 1–13.

—— 1981: Preface. In K. R. Cox and R. G. Golledge (eds.), *Behavioral problems in geography revisited* (London: Methuen), xiii–xxix.

COX, K. R. and McCARTHY, J. J. 1982: Neighbourhood activism as a politics of turf: a critical analysis. In K. R. Cox and R. J. Johnston, (eds.), *Conflict, politics and the urban scene* (London: Longman), 196–219.

COX, K. R. REYNOLDS, D. R. and ROKKAN, S. (eds.) 1974: *Locational approaches to power and conflict*. New York: Halsted Press.

CRANE, D. 1972: *Invisible colleges*. Chicago: University of Chicago Press.

CROWE, P. R. 1936: The rainfall regime of the Western Plains. *Geographical Review* **26**, 463–84.

—— 1938: On progress in geography. *Scottish Geographical Magazine* **54**, 1–19.

—— 1970: Review of *Progress in Geography* 1. *Geography* **55**, 346–7.

CUMBERLAND, K. B. 1947: *Soil erosion in New Zealand*. Wellington: Whitcombe & Tombs.

CURRY, L. 1967: Quantitative geography. *The Canadian Geographer* **11**, 265–74.

—— 1972: A spatial analysis of gravity flows. *Regional Studies* **6**, 131–47.

DACEY, M.F. 1962: Analysis of central-place and point patterns by a nearest-neighbor method. In K. Norborg (ed.), *Proceedings of the IGU Symposium in Urban Geography, Lund 1960* (Lund: C. W. K. Gleerup), 55–76.

—— 1968: A review on measures of contiguity for two and k-color maps. In B. J. L. Berry and D. F. Marble (eds.), *Spatial analysis* (Englewood Cliffs, NJ: Prentice-Hall), 479–95.

—— 1973: Some questions about spatial distributions. In R. J. Chorley (ed.), *Directions in geography* (London: Methuen), 127–52.

DARBY, H. C. 1953: On the relations of geography and history. *Transactions and Papers, Institute of British Geographers* **19**, 1–11.

—— 1962: The problem of geographical description. *Transactions and Papers, Institute of British Geographers* **30**, 1–14.

—— (ed.) 1973: *A new historical geography of England*. London: Cambridge University Press.

—— 1977: *Domesday England*. London: Cambridge University Press.

DAVIES, W. K. D. 1972a: Geography and the methods of modern science, In W. K. D. Davies (ed.), *The conceptual revolution in geography* (London: University of London Press), 131–9.

—— 1972b: Introduction: the conceptual revolution in geography. In W. K. D. Davies (ed.) *The conceptual revolution in geography* (London: University of London Press), 9–18.

DAVIS, W. M. 1906: An inductive study of the content of geography. *Bulletin of the American Geographical Society* **38**, 67–84.

DAY, M. and TIVERS, J. 1979: Catastrophe theory and geography: a Marxist critique. *Area* **11**, 54–8.

DAYSH, G. H. J. (ed.) 1949: *Studies in regional planning*. London: Philip & Son.

DEAR, M. and CLARK, G. L. 1978: The state and geographic process: a critical review. *Environment and Planning, A* **10**, 173–84.

DEAR, M. and SCOTT, A. J. (eds.) 1981: *Urbanization and urban planning in capitalist society*. London: Methuen.

DETWYLER, T. R. and MARCUS, M. G. (eds.) 1972: *Urbanization and environment*. Belmont, California: Duxbury Press.

DICKENSON, J. P. and CLARKE, C. G. 1972: Relevance and the 'newest geography'. *Area* **3**, 25–7.

DICKINSON, R. E. 1973: The distribution and functions of smaller urban settlements of East Anglia. *Geography* **18**, 19–31.

—— 1947: *City, region and regionalism.* London: Routledge & Kegan Paul.

DOWNS, R. M. 1970: Geographic space perception: past approaches and future prospects. In C. Board *et al.* (eds.), *Progress in Geography* **2** (London: Edward Arnold), 65–108.

—— 1979: Critical appraisal or determined philosophical skepticism? *Annals, Association of American Geographers* **69**, 468–71.

DOWNS, R. M. and MEYER, J. T. 1978: Geography and the mind. *Human geography: coming of age. American Behavioral Scientist* **22**, 59–78.

DOWNS, R. M. and STEA, D. (eds.) 1973: *Image and environment.* London: Edward Arnold.

—— 1977: *Maps in mind.* New York: Harper & Row.

DRYSDALE, A. and WATTS, M. 1977: Modernization and social protest movements. *Antipode* **9**(1), 40–55.

DUNCAN, J. S. 1980: The superorganic in American cultural geography. *Annals, Association of American Geographers* **70**, 181–98.

DUNCAN, J. S. and LEY, D. 1982: Structural marxism and human geography: a critical assessment. *Annals, Association of American Geographers* **72**, 30–59.

DUNCAN, O. D. 1959: Human ecology and population studies. In P. M. Hauser and O. D. Duncan (eds.), *The study of population* (Chicago: University of Chicago Press), 678–716.

DUNCAN, O. D., CUZZORT, R. P. and DUNCAN, B. 1961: *Statistical geography.* New York: The Free Press.

DUNCAN, O. D. and SCHNORE, L. F. 1959: Cultural, behavioral and ecological perspectives in the study of social organization. *American Journal of Sociology* **65**, 132–46.

DUNCAN, S. S. 1974a: Cosmetic planning or social engineering? Improvement grants and improvement areas in Huddersfield. *Area* **6**, 259–70.

—— 1974b: The isolation of scientific discovery: indifference and resistance to a new idea. *Science Studies* **4**, 109–34.

—— 1975: Research directions in social geography: housing opportunities and constraints. *Transactions, Institute of British Geographers* **NS1** 10–19.

—— 1979: Radical geography and marxism. *Area* **11**, 124–6.

—— 1981: Housing policy, the methodology of levels, and urban research: the case of Castells. *International Journal of Urban and Regional Research* **5**, 231–54.

DUNFORD, M. F. 1982: *Capital and the space-economy.* London: Macmillan.

EILON, S. 1975: Seven faces of research. *Operational Research Quarterly* **26**, 359–67.

ELIOT HURST, M. E. 1972: Establishment geography: or how to be irrelevant in three easy lessons. *Antipode* **5**(2), 40–59.

—— 1973: Transportation and the societal framework. *Economic Geography* **49**, 163–80.

—— 1980: Geography social science and society: towards a de-definition. *Australian Geographical Studies* **18**, 3–21.

ENTRIKIN, J. N. 1976: Comtemporary humanism in geography. *Annals, Association of American Geographers* **66**, 615–32.

—— 1980: Robert Park's human ecology and human geography *Annals, Association of American Geographers* **70**, 615–32.

—— 1981: Philosophical issues in the scientific study of regions. In D. T. Herbert and R. J. Johnston (eds.), *Geography and the urban environment,* volume 4 (Chichester: John Wiley), 1–27.

EYLES, J. 1971: Pouring new sentiments into old theories: how else can we look at behavioural patterns? *Area* **3**, 242–50.
—— 1973: Geography and relevance. *Area* **5**, 158–60.
—— 1974: Social theory and social geography. In C. Board *et al.* (eds.), *Progress in Geography* **6** (London: Edward Arnold), 27–88.
—— 1981: Why geography cannot be Marxist: towards an understanding of lived experience. *Environment and Planning* **A 13**, 1371–88.
EYLES, J. and LEE, R. 1982: Human geography in explanation. *Transactions, Institute of British Geographers* **NS7**, 117–12.
EYLES, J. D. and SMITH, D. M. 1982: Social geography. *Human geography: Coming of Age. American Behavioral Scientist* **22**, 41–58.
EYRE, S. R. 1978: *The real wealth of nations*. London: Edward Arnold.
EYRE, S. R. and JONES, G. R. J. (ed.) 1666: *Geography as human ecology*. London: Edward Arnold.
FEYERABEND, P. 1975: *Against method*. London: New Left Books.
FLEURE, H. J. 1919: Human regions. *Scottish Geographical Magazine* **35**, 94–105.
FOLKE, S. 1972: Why a radical geography must be Marxist. *Antipode* **4**(2), 13–18.
—— 1973: First thoughts on the geography of imperialism. *Antipode* **5**(3), 16–20.
FOOTE, D. C. and GREER-WOOTTEN, B. 1968: An approach to systems analysis in cultural geography. *The Professional Geographer* **20**, 86–90.
FORD, L. R. 1982: Beware of new geographies. *The Professional Geographer* **34**, 131–5.
FORER, P. C. 1974: Space through time: a case study with New Zealand airlines. In E. L. Cripps (ed.), *Space-Time Concepts in Urban and Regional Models* (London: Pion), 22–45.
FORRESTER, J. W. 1969: *Urban dynamics*. Cambridge, Mass.: MIT Press.
FOTHERINGHAM, A. S. 1981: Spatial structure and distance-decay parameters. *Annals, Association of American Geographers* **71**, 425–36.
FOUCAULT, M. 1972: *The archaeology of knowledge*. London: Tavistock Publications.
FREEMAN, T. W. 1961: *A hundred years of geography*. London: Gerald Duckworth.
—— 1979: The British school of geography. *Organon* **14**, 205–16.
—— 1980a: *A history of modern British geography*. London: Longman.
—— 1980b: The Royal Geographical Society and the development of geography. In E. H. Brown (ed.), *Geography, yesterday and tomorrow* (Oxford: Oxford University Press), 1–99.
FULLER, G. A. 1971: The geography of prophylaxis: an example of intuitive schemes and spatial competition in Latin America. *Antipode* **3**(1), 21–30.
GALE, N. and GOLLEDGE, R. G. 1982: On the subjective partitioning of space *Annals, Association of American Geographers* **72**, 60–7.
GARRISON, W. L. 1953: Remoteness and the passenger utilization of air transportation. *Annals, Association of American Geographers* **43**, 169.
—— 1956a: Applicability of statistical inference to geographical research. *Geographical Review* **46**, 427–9.
—— 1956b: Some confusing aspects of common measurements. *The Professional Geographer* **8**, 4–5.
—— 1959a: Spatial structure of the economy I. *Annals, Association of American Geographers* **49**, 238–9.

—— 1959b: Spatial structure of the economy II. *Annals, Association of American Geographers* **49**, 471–82.

—— 1960a: Spatial structure of the economy III. *Annals, Association of American Geographers* **50**, 357–73.

—— 1960b: Connectivity of the interstate highway system. *Papers and Proceedings, Regional Science Association* **6**, 121–37.

—— 1962: Simulation models of urban growth and development. In K. Norborg (ed.), *IGU Symposium in Urban Geography*, Lund Studies in Geography B 24 (Lund: C. W. K. Gleerup), 91–108.

—— 1979: Playing with ideas. *Annals, Association of American Geographers* **69**, 118–20.

GARRISON, W. L., BERRY, B. J. L., MARBLE, D. F., NYSTUEN, J. D. and MORRILL, R. L. 1959: *Studies of highway development and geographic change.* Seattle: University of Washington Press.

GARRISON, W. L. and MARBLE, D. F. 1957: The spatial structure of agricultural activities. *Annals, Association of American Geographers* **47**, 137–44.

—— (eds.), 1967a: *Quantitative Geography. Part I: Economic and Cultural Topics.* Northwestern University Studies in Geography, Number 13, Evanston, Illinois.

—— (eds.) 1967b: *Quantitative Geography, Part II: Physical and Cartographic Topics.* Northwestern University Studies in Geography, Number 14, Evanston, Illinois.

GATRELL, A. C. 1982: *Geometry in geography and the geometry of geography.* Discussion Paper 6, Department of Geography, University of Salford.

GEARY, R. C. 1954: The contiguity ratio and statistical mapping. *The Incorporated Statistician* **5**, 115–41.

GETIS, A. 1963: The determination of the location of retail activities with the use of a map transformation. *Economic Geography* **39**, 1–22.

GETIS, A. and BOOT, B. N. 1978: *Models of spatial processes.* London: Cambridge University Press.

GIBSON, E. 1978: Understanding the subjective meaning of places. In D. Ley and M. S. Samuels (eds.), *Humanistic geography: problems and prospects* (Chicago: Maaroufa Press), 138–54.

GIDDENS, A. 1981: *A critique of contemporary historical materialism.* London: Macmillan.

GINSBURG, N. 1972: The mission of a scholarly society. *The Professional Geographer* **24**, 1–6.

—— 1973: From colonialism to national development: geographical perspectives on patterns and policies. *Annals, Association of American Geographers* **63**, 1–21.

GLACKEN, C. J. 1956: Changing ideas of the habitable world. In W. L. Thomas (ed.), *Man's role in changing the face of the earth* (Chicago: University of Chicago Press), 70–92.

—— 1967: *Traces on the Rhodian shore: nature and culture in western thought from ancient times to the end of the eighteenth century.* Berkeley: University of California Press.

GOHEEN, P. G. 1970: *Victorian Toronto.* University of Chicago, Department of Geography, Research Paper 127.

GOLD, J. R. 1980: *An introduction to behavioural geography.* Oxford: Oxford University Press.

GOLLEDGE, R. G. 1969: The geographical relevance of some learning theories. In

K. R. Cox and R. G. Golledge (eds.), *Behavioral problems in geography: a symposium* (Evanston: Northwestern University Studies in Geography 17), 101–45.

—— 1970: Some equilibrium models of consumer behavior. *Economic Geography* **46**, 417–24.

—— 1980: A behavioral view of mobility and migration research. *The Professional Geographer* **32**, 14–21.

—— 1981a: Misconceptions, misinterpretations, and misrepresentations of behavioral approaches in human geography. *Environment and Planning A* **13**, 1325–44.

—— 1981b: A critical response to Guelke's 'Uncritical rhetoric'. *The Professional Geographer* **33**, 247–51.

GOLLEDGE, R. G. and AMEDEO, D. 1968: On laws in geography. *Annals, Association of American Geography* **58**, 760–74.

GOLLEDGE, R. G. and BROWN, L. A. 1967: Search, learning and the market decision process. *Geografiska Annaler* **49B**, 116–24.

GOLLEDGE, R. G., BROWN, L. A. and WILLIAMSON, F. 1972: Behavioral approaches in geography: an overview. *The Australian Geographer* **12**, 59–79.

GOULD, P. R. 1963: Man against his environment: a game theoretic framework. *Annals, Association of American Geographers* **53**, 290–7.

—— 1966: On mental maps. Michigan Inter-University Community of Mathematical Geographers, Discussion Paper 9. Reprinted in R. M. Downs and D. Stea, 1973, *Image and environment* (London: Edward Arnold), 182–220.

—— 1969: Methodological developments since the fifties. In C. Board *et al.* (eds.) *Progress in Geography* **1**, 1–50.

—— 1970a: Is *statistix inferens* the geographical name for a wild goose? *Economic Geography* **46**, 439–48.

—— 1970b: Tanzania 1920–63: the spatial impress of the modernization process. *World Politics* **22**, 149–70.

—— 1972: Pedagogic review. *Annals, Association of American Geographers* **62**, 689–700.

—— 1975: Mathematics in geography: conceptual revolution or new tool? *International Social Science Journal* **27**, 303–27.

—— 1977: What is worth teaching in geography? *Journal of Geography in Higher Education* **1**, 20–36.

—— 1978: Concerning a geographic education. In D. A. Lanegran and R. Palm (eds.), *An invitation to geography* (New York: McGraw Hill), 202–26.

—— 1979: Geography 1957–1977: the Augean period. *Annals, Association of American Geographers* **69**, 139–51.

—— 1981: Letting the data speak for themselves. *Annals, Association of American Geographers* **71**, 166–76.

GOULD, P. R. and WHITE, R. 1974: *Mental maps*. Harmondsworth: Penguin Books.

GRANO, O. 1981: External influence and internal change in the development of geography. In D. R. Stoddart (ed.), *Geography, ideology and social concern* (Oxford: Blackwell), 17–36.

GRAVES, N. J. 1981: Can geographical studies be subsumed under one paradigm or are a plurality of paradigms inevitable? *Terra* **93**, 85–90.

GRAY, F. 1975: Non-explanation in urban geography. *Area* **7**, 228–35.

—— 1976: Selection and allocation in council housing. *Transactions, Institute of British Geographers* **NS1**, 34–46.

GREER-WOOTTEN, B. 1972: *The role of general systems theory in geographic research*. Department of Geography, York University, Discussion Paper No. 3, Toronto.

GREGORY, D. 1976: Rethinking historical geography. *Area* **8**, 295–9.

— 1978a: *Ideology, science and human geography*. London: Hutchinson.

— 1978b: The discourse of the past: phenomenology, structuralism, and historical geography. *Journal of Historical Geography* **4**, 161–73.

— 1980: The ideology of control: systems theory and geography. *Tijdschrift voor Economische en Sociale Geografie* **71**, 327–42.

— 1981: Human agency and human geography. *Transactions, Institute of British Geographers* **NS6**, 1–18.

— 1982a: *Regional transformation and industrial revolution: a geography of the Yorkshire woollen industry*. London: Macmillan.

— 1982b: Solid geometry: notes on the recovery of spatial structure. In P. R. Gould and G. Olsson (eds.), *A search for common ground* (London: Pion), 187–222.

GREGORY, S. 1963: *Statistical methods and the geographer*. London: Longman.

— 1976: On geographical myths and statistical fables. *Transactions, Institute of British Geographers* **NS1**, 385–400.

GRIGG, D. B. 1977: E. G. Ravenstein and the laws of migration. *Journal of Historical Geography* **3**, 41–54.

GROSSMAN, L. 1977: Man-environment relationships in anthropology and geography. *Annals, Association of American Geographers* **67**, 126–44.

GUDGIN, G. and TAYLOR, P. J. 1979: *Seats, votes and the spatial organisation of elections*. London: Pion.

GUELKE, L. 1971: Problems of scientific explanation in geography. *The Canadian Geographer* **15**, 38–53.

— 1974: An idealist alternative in human geography. *Annals, Association of American Geographers* **14**, 193–202.

— 1975: On rethinking historical geography. *Area* **7**, 135–8.

— 1976: The philosophy of idealism. *Annals, Association of American Geography* **66**, 168–9.

— 1977a: The role of laws in human geography. *Progress in Human Geography* **1**, 376–86.

— 1977b: Regional geography. *The Professional Geographer* **29**, 1–7.

— 1978: Geography and logical positivism. In D. T. Herbert and R. J. Johnston (eds.), *Geography and the urban environment: progress in research and applications*, volume 1 (London: John Wiley), 35–61.

— 1981a: Uncritical rhetoric: 'A classic disservice'. *The Professional Geographer* **33**, 246–7.

— 1981b: Idealism. In M. E. Harvey and B. P. Holly (eds.), *Themes in geographic thought* (London: Croom Helm), 133–47.

GUTTING, G. 1980: Introduction. In G. Gutting (ed.), *Paradigms and revolutions* (Notre Dame, Indiana: University of Notre Dame Press), 1–22.

HÄGERSTRAND, T. 1968: *Innovation diffusion as a spatial process*. Chicago: University of Chicago Press.

— 1977: The geographers' contribution to regional policy: the case of Sweden. In D. R. Deskins *et al.* (eds.), *Geographic humanism, analysis and social action: a half century of geography at Michigan* (Michigan Geographical Publications No. 17, Ann Arbor), 329–46.

HAGGETT, P. 1964: Regional and local components in the distribution of forested areas in southeast Brazil: a multivariate approach. *Geographical Journal* **130**, 365–77.

— 1965a: Changing concepts in economic geography. In R. J. Chorley and P. Haggett (eds.), *Frontiers in geographical teaching* (London: Methuen), 101–17.

— 1965b: Scale components in geographical problems. In R. J. Chorley and P. Haggett (eds.), *Frontiers in geographical teaching* (London: Methuen), 164–85.

— 1965c: *Locational analysis in human geography*. London: Edward Arnold.

— 1967: Network models in geography. In R. J. Chorley and P. Haggett (eds.) *Models in geography* (London: Methuen), 609–70.

— 1973: Forecasting alternative spatial, ecological and regional futures: problems and possibilities. In R. J. Chorley (ed.), *Directions in geography* (London: Methuen), 217–36.

— 1978: The spatial economy. *Human Geography: Coming of Age. American Behavioral Scientist* **22**, 151–67.

HAGGETT, P. and CHORLEY, R. J. 1965: Frontier movements and the geographical tradition. In R. J. Chorley and P. Haggett (eds.), *Frontiers in geographical teaching* (London: Methuen) 358–78.

— 1967: Models, paradigms, and the new geography. In R. J. Chorley and P. Haggett (eds.), *Models in geography* (London: Methuen), 19–42.

— 1969: *Network models in geography*. London: Edward Arnold.

HAGGETT, P. CLIFF, A. D. and FREY, A. 1977: *Locational analysis in human geography*. London: Edward Arnold.

HAGOOD, M. J. 1943: Development of a 1940 rural farm level of living index for counties. *Rural Sociology* **8**, 171–80.

HAINING, R. P. 1980: Spatial autocorrelation problems. In D. T. Herbert and R. J. Johnston (eds.), *Geography and the urban environment*, volume 3 (Chichester: John Wiley), 1–44.

— 1981: Analysing univariate maps, *Progress in Human Geography* **5**, 58–78.

HALL, P. 1974: The new political geography. *Transactions, Institute of British Geographers* **63**, 48–52.

— 1981a: *Great planning disasters*. London: Penguin.

— 1981b: The geographer and society. *Geographical Journal* **147**, 145–52.

— 1982: The new political geography: seven years on, *Political Geography Quarterly* **1**, 65–76.

HALVORSON, P. and STAVE, B. M. 1978: A conversation with Brian J. L. Berry. *Journal of Urban History* **4**, 209–38.

HAMILTON, F. E. I. 1974: A view of spatial behaviour, industrial organizations, and decision-making. In F. E. I. Hamilton (ed.), *Spatial perspectives on industrial organization and decision-making* (London: John Wiley), 3–46.

HAMNETT, C. 1977: Non-explanation in urban geography: throwing the baby out with the bath water. *Area* **9**, 143–5.

HARE, F. K. 1974: Geography and public policy: a Canadian view. *Transactions, Institute of British Geographers* **63**, 25–8.

— 1977: Man's world and geographers: a secular sermon. In D. R. Deskins *et al.* (eds.), *Geographic humanism, analysis and social action: a half century of geography at Michigan* (Michigan Geographical Publication No. 17, Ann Arbor), 259–73.

HARRIES, K. D. 1974: *The geography of crime and justice*. New York: McGraw Hill.

—— 1975: Rejoinder to Richard Peet: 'The geography of crime: a political critique.' *The Professional Geographer* **27**, 280-2.

—— 1976: Observations on radical versus liberal theories of crime causation. *The Professional Geographer* **28**, 100-13.

HARRIS, C. D. 1954a: The geography of manufacturing. In P. E. James and C. F. Jones (eds.), *American geography: inventory and prospect* (Syracuse: Syracuse University Press), 292-309.

—— 1954b: The market as a factor in the localization of industry in the United States. *Annals, Association of American Geographers* **44**, 315-48.

—— 1977: Edward Louis Ullman, 1912-1976. *Annals, Association of American Geographers* **67**, 595-600.

HARRIS, C. D. and ULLMAN, E. L. 1945: The nature of cities. *Annals of the American Academy of Political and Social Science* **242**, 7-17.

HARRIS, R. C. 1971: Theory and synthesis in historical geography. *The Canadian Geographer* **15**, 157-72.

—— 1977: The simplification of Europe overseas. *Annals, Association of American Geographers* **67**, 469-83.

—— 1978: The historical mind and the practice of geography. In D. Ley and M. S. Samuels (eds.), *Humanistic geography: problems and prospects* (Chicago: Maaroufa Press), 123-37.

HARRISON, R. T. and LIVINGSTONE, D. N. 1982: Understanding in geography: structuring the subjective. In D. T. Herbert and R. J. Johnston (eds.), *Geography and the urban environment*, volume 5 (Chichester: John Wiley), 1-40.

HART, J. F. 1982: The highest form of the geographer's art. *Annals, Association of American Geographers* **72**, 1-29.

HARTSHORNE, R. 1939: *The nature of geography*. Lancaster, Pennsylvania: Association of American Geographers.

—— 1948: On the mores of methodological discussion in American geography. *Annals, Association of American Geographers* **38**, 492-504.

—— 1954a: Political geography. In P. E. James and C. F. Jones (eds.), *American geography: inventory and prospect* (Syracuse: Syracuse University Press), 167-225.

—— 1954b: Comment on 'Exceptionalism in geography'. *Annals, Association of American Geographers* **44**, 108-9.

—— 1955: 'Exceptionalism in geography' re-examined. *Annals, Association of American Geographers* **45**, 205-44.

—— 1958: The concept of geography as a science of space from Kant and Humboldt to Hettner. *Annals, Association of American Geographers* **48**, 97-108.

—— 1959: *Perspective on the nature of geography*. Chicago: Rand McNally.

—— 1972: Review of *Kant's concept of geography*. *The Canadian Geographer* **16**, 77-9.

—— 1979: Notes towards a bibliography of the nature of geography. *Annals, Association of American Geographers* **69**, 63-76.

HARVEY, D. 1967a: Models of the evolution of spatial patterns in geography. In R. J. Chorley and P. Haggett (eds.), *Models in Geography* (London: Methuen), 549-608.

—— 1967b: Editorial introduction: the problem of theory construction in geography. *Journal of Regional Science* **7**, 211-16.

—— 1969a: *Explanation in geography*. London: Edward Arnold.

—— 1969b: Review of A. Pred, *Behavior and location part I. Geographical Review* **59**, 312–14.

—— 1969c: Conceptual and measurement problems in the cognitive-behavioral approach to location theory. In K. R. Cox and R. G. Golledge (eds.), *Behavioral problems in geography: a symposium* (Northwestern University Studies in Geography 17), 35–68.

—— 1970: Behavioral postulates and the construction of theory in human geography. *Geographica Polonica* **18**, 27–46.

—— 1972: Revolutionary and counter-revolutionary theory in geography and the problem of ghetto formation. *Antipode* **4**(2), 1–13.

—— 1973: *Social justice and the city*. London: Edward Arnold.

—— 1974a: A commentary on the comments. *Antipode* **4**(2), 36–41.

—— 1974b: Discussion with Brian Berry. *Antipode* **6**(2), 145–8.

—— 1974c: What kind of geography for what kind of public policy? *Transactions, Institute of British Geographers* **63**, 18–24.

—— 1974d: Population, resources and the ideology of science. *Economic Geography* **50**, 256–77.

—— 1974e: Class-monopoly rent, finance capital and the urban revolution, *Regional Studies* **8**, 239–55.

—— 1975a: Class structure in a capitalist society and the theory of residential differentiation. In R. Peel, M. Chisholm and P. Haggett (eds.), *Processes in physical and human geography: Bristol essays* (London: Heinemann), 354–69.

—— 1975b: The political economy of urbanization in advanced capitalist societies: the case of the United States. In G. Gappert and H. M. Rose (eds.), *The social economy of cities* (Beverly Hills: Sage Publications), 119–63.

—— 1975c: Review of B. J. L. Berry, *The human consequences of urbanization. Annals, Association of American Geographers* **65**, 99–103.

—— 1976: The marxist theory of the state. *Antipode* **8**(2) 80–9.

—— 1978: The urban process under capitalism: a framework for analysis. *International Journal of Urban and Regional Research* **2**, 101–32.

—— 1982: *The limits to capital*. Oxford: Blackwell.

HARVEY, M. E. and HOLLY, B. P. 1981: Paradigm, philosophy and geographic thought. In M. E. Harvey and B. P. Holly (eds.), *Themes in geographic thought* (London: Croom Helm), 11–37.

HAY, A. M. 1978: Some problems in regional forecasting. In J. I. Clarke and J. Pelletser (eds.), *Regions geographique et regions d'amenagement*. Collection les hommes et les lettres, 7. Lyon: Editions Hermes.

—— 1979a: Positivism in human geography: response to critics. In D. T. Herbert and R. J. Johnston (eds.), *Geography and the urban environment: progress in research and applications*, volume 2 (London: John Wiley), 1–26.

—— 1979b: The geographical explanation of commodity flow. *Progress in Human Geography* **3**, 1–12.

HAYNES, R. M. 1975: Dimensional analysis: some applications in human geography. *Geographical Analysis* **7**, 51–68.

—— 1978: A note on dimensions and relationships in human geography. *Geographical Analysis* **10**, 288–92.

—— 1982: *An introduction to dimensional analysis for geographers*. CATMOG 33, Geo Abstracts, Norwich.

HAYTER, R. and WATTS, H. D. 1983: The geography of enterprise. *Progress in Human Geography* **7**.

HERBERT, D. T. and JOHNSTON, R. J. 1978: Geography and the urban environment. In D. T. Herbert and R. J. Johnston (eds.), *Geography and the urban environment: progress in research and applications,* volume 1 (London: John Wiley), 1–29.

HERBERTSON, A. J. 1905: The major natural regions, *Geographical Journal* **25**, 300–10.

HODGART, R. L. 1978: Optimizing access to public services: a review of problems, models, and methods of locating central facilities. *Progress in Human Geography* **2**, 17–48.

HOLT-JENSEN, A. 1981: *Geography: its history and concepts.* London: Harper & Row.

HOOK, J. C. 1955: Areal differentiation of the density of the rural farm population in the northeastern United States. *Annals, Association of American Geographers* **45**, 189–90.

HOOSON, D. J. M. 1981: Carl O. Sauer. In B. W. Blouet (ed.), *The origins of academic geography in the United States* (Hamden, Conn: Archon Books), 165–74.

HOUSE, J. W. 1973: Geographers, decision takers and policy makers. In M. Chisholm and B. Rodgers (eds.), *Studies in human geography* (London: Heinemann), 272–305.

HUGGETT, R. J. 1980: *Systems analysis in geography.* Oxford: Oxford University Press.

HUGGETT, R. J. and THOMAS, R. W. 1980: *Modelling in geography.* London: Harper & Row.

ISARD, W. 1956a: *Location and space economy.* New York: John Wiley.

—— 1956b: Regional science, the concept of region, and regional structure. *Papers and Proceedings, Regional Science Association* **2**, 13–39.

—— 1960: *Methods of regional analysis: an introduction in regional science.* New York: John Wiley.

ISARD, W. *et al.* 1969: *General theory: social, political, economic and regional.* Cambridge, Mass: The MIT Press.

ISARD, W. 1975: *An introduction to regional science.* Englewood Cliffs, NJ: Prentice-Hall.

JACKSON, P. and SMITH, S. J. 1981: Introduction. In P. Jackson and S. J. Smith (eds.), *Social interaction and ethnic segregation* (London: Academic Press), 1–18.

JAMES, P. E. 1942: *Latin America.* London: Cassell.

—— 1954: Introduction: the field of geography. in P. E. James and C. F. Jones (eds.), *American geography: inventory and prospect* (Syracuse: Syracuse University Press), 2–18.

—— 1965: The President's session. *The Professional Geographer* **17**(4), 35–7.

—— 1972: *All possible worlds: a history of geographical ideas.* Indianapolis: The Odyssey Press.

JAMES, P. E. and JONES, C. F. (eds.) 1954: *American geography: inventory and prospect.* Syracuse: Syracuse University Press.

JAMES, P. E. and MARTIN, G. J. 1981: *All possible worlds: a history of geographical ideas* (2nd edn). New York: John Wiley.

JANELLE, D. G. 1968: Central-place development in a time-space framework. *The Professional Geographer* **20**, 5–10.

—— 1969: Spatial reorganization: a model and concept. *Annals, Association of American Geographers* **59**, 348–64.

JOHNSON, J. H. and POOLEY, C. G. (eds.) 1982: *The structure of nineteenth-century cities*. London: Croom Helm.

JOHNSTON, R. J. 1969: Urban geography in New Zealand 1945–1969. *New Zealand Geographer* **25**, 121–35.

—— 1971: *Urban residential patterns: an introductory review*. London: G. Bell & Sons.

—— 1974: Continually changing human geography revisited: David Harvey: *Social Justice and the City. New Zealand Geographer* **30**, 180–92.

—— 1976a: *The world trade system: some enquiries into its spatial structure*. London: G. Bell & Sons.

—— 1976b: Anarchy, conspiracy and apathy: the three 'conditions' of geography. *Area* **8**, 1–3.

—— 1978a: *Multivariate statistical analysis in geography: a primer on the general linear model*. London: Longman.

—— 1978b: *Political, electoral and spatial systems*. London: Oxford University Press.

—— 1978c: Paradigms and revolutions or evolution: observations on human geography since the Second World War. *Progress in Human Geography* **2**, 189–206.

—— 1979a: Urban geography: city structures. *Progress in Human Geography* **3**, 133–8.

—— 1979b: *Geography and geographers: Anglo-American human geography since 1945* (1st edn), London: Edward Arnold.

—— 1980a: On the nature of explanation in human geography. *Transactions, Institute of British Geographers* **NS5**, 402–12.

—— 1980b: *City and society*. London: Penguin.

—— 1981a: Applied geography, quantitative analysis and ideology. *Applied Geography* **1**, 213–9.

—— 1981b: Paradigms, revolutions, schools of thought and anarchy: reflections on the recent history of Anglo-American human geography. In B. W. Blouet (ed.), *The origins of academic geography in the United States* (Hamden, Conn.: Archon Books), 303–18.

—— 1982a: *Geography and the state*. London: Macmillan.

—— 1982b: A foundling floundering in World Three. In M. Billinge, D. Gregory and R. L. Martin (eds.), *Reflections on a revolution* (London: Macmillan).

—— 1982c: On the nature of human geography. *Transactions, Institute of British Geographers* **NS7**, 123–5.

—— 1982d: On ecological analysis and spatial autocorrelation. In L. le Rouzic, (ed.), *L'autocorrelation spatiale* (Reims, Travaux de l'Institute de Geographie), 3–16.

—— 1983a: Resource analysis, resource management and the integration of human and physical geography. *Progress in Physical Geography*, **7**.

—— 1983b: *Philosophy and human geography: an introduction to contemporary approaches*. London: Edward Arnold.

—— 1983c: Texts, actors, and higher managers: judges, bureaucrats and the political organization of space, *Political Geography Quarterly* **2**.

JOHNSTON, R. J. and CLAVAL, P. (eds.) 1983: *Geography since the Second World War: an international survey*. London: Croom Helm.

JOHNSTON, R. J. and GREGORY, S. 1983: The United Kingdom. In R. J. Johnston and P. Claval (eds.), *Geography since the Second World War: an international survey* (London: Croom Helm).

JOHNSTON, R. J. and HERBERT, D. T. 1978: Introduction. In D. T. Herbert and R. J. Johnston (eds.), *Social areas in cities: processes, patterns and problems* (London: John Wiley), 1–33.

JONES, E. 1956: Cause and effect in human geography. *Annals, Association of American Geographers* **46**, 369–77.

—— 1980: Social geography. In E. H. Brown, (ed.), *Geography, yesterday and tomorrow* (Oxford: Oxford University Press), 251–62.

KANSKY, K. J. 1963: *Structure of transport networks.* University of Chicago, Department of Geography, Research Paper 84, Chicago.

KASPERSON, R. E. 1971: The post-behavioral revolution geography. *British Columbia Geographical Series* **12**, 5–20.

KATES, R. W. 1962: *Hazard and choice perception in flood plain management.* University of Chicago, Department of Geography, Research Paper 78, Chicago.

—— 1972: Review of *Perspectives on resource management. Annals, Association of American Geographers* **62**, 519–20.

KEEBLE, D. E. 1976: *Industrial location and planning in the United Kingdom.* London: Methuen.

KING, L. J. 1960: A note on theory and reality. *The Professional Geographer* **12**(3), 4–6.

—— 1961: A multivariate analysis of the spacing of urban settlement in the United States. *Annals, Association of American Geographers* **51**, 222–3.

—— 1969: The analysis of spatial form and relationship to geographic theory. *Annals, Association of American Geographers* **59**, 573–95.

—— 1976: Alternatives to a positive economic geography. *Annals, Association of American Geographers* **66**, 293–308.

—— 1979a: Areal associations and regressions. *Annals, Association of American Geographers* **69**, 124–8.

—— 1979b: The seventies: disillusionment and consolidation. *Annals, Association of American Geographers* **69**, 155–7.

KING, L. J. and CLARK, G. L. 1978: Government policy and regional development. *Progress in Human Geography* **2**, 1–16.

KIRK, W. 1951: Historical geography and the concept of the behavioural environment. *Indian Geographical Journal* **25**, 152–60.

—— 1963: Problems of geography. *Geography* **48**, 357–71.

—— 1978: The road from Mandalay: towards a geographical philosophy. *Transactions, Institute of British Geographers* **NS3**, 381–394.

KISH, G. and WARD, R. 1981: A survival package for geography and other endangered disciplines. *Newsletter*, Association of American Geographers, **16**, pp. 8,14.

KNOS, D. S. 1968: The distribution of land values in Topeka, Kansas. In B. J. L. Berry and D. F. Marble (eds.), *Spatial analysis* (Englewood Cliffs, NJ: Prentice-Hall), 269–89.

KNOX, P. L. 1975: *Social well-being: a spatial perspective.* London: Oxford University Press.

KOLLMORGEN, W. N. 1979: Kollmorgen as a bureaucrat. *Annals, Association of American Geographers* **69**, 77–89.

KUHN, T. S. 1962: *The structure of scientific revolutions.* Chicago: University of Chicago Press.

—— 1970a: *The structure of scientific revolutions* (2nd edn). Chicago: University of Chicago Press.

—— 1970b: Logic of discovery or psychology of research? In I. Lakatos and A. Musgrave (eds.), *Criticism and the growth of knowledge* (Cambridge: Cambridge University Press), 1–23.

—— 1970c: Reflections on my critics. In I. Lakatos and A. Musgrave (eds.), *Criticism and the growth of knowledge* (Cambridge: Cambridge University Press), 231–78.

—— 1977: Second thoughts on paradigms. In F. Suppe (ed.), *The structure of scientific theories* (Urbana: University of Illinois Press), 459–82, plus discussion 500–17.

LABEDZ, L. 1977: Anarchism. In A. Bullock and O. Stallybrass (eds.), *The Fontana dictionary of modern thought* (London: Fontana), 22.

LAKATOS, I. 1978: Falsification and the methodology of scientific research programmes. In J. Worrall and G. Currie (eds.), *The methodology of scientific research programmes, philosophical papers*, volume I (Cambridge: Cambridge University Press), 8–101.

LANGTON, J. 1972: Potentialities and problems of adapting a systems approach to the study of change in human geography. *Progress in Geography* **4**, 125–79.

LAPONCE, J. A. 1980: Political science: an import-export analysis of journals and footnotes. *Political Studies* **28**, 401–19.

LAVALLE, P., McCONNELL, H. and BROWN, R. G. 1967: Certain aspects of the expansion of quantitative methodology in American geography. *Annals, Association of American Geographers* **57**, 423–36.

LAW, J. 1976: Theories and methods in the sociology of science: an interpretative approach. In G. Lemaine *et al.*, *Perspectives on the emergence of scientific disciplines* (The Hague: Mouton), 221–31.

LEACH, B. 1974: Race, problems and geography. *Transactions, Institute of British Geographers* **63**, 41–7.

LEACH, E. R. 1974: *Lévi-Strauss*. London: Fontana.

LEE, Y. 1975: A rejoinder to 'The geography of crime: a political critique'. *The Professional Geographer* **27**, 284–5.

LEMAINE, G., *et al.* 1976: Introduction: problems in the emergence of new disciplines. In G. Lemaine *et al.* (eds.), *Perspectives on the emergence of scientific disciplines* (The Hague: Mouton), 1–73.

LEONARD, S. 1982: Urban managerialism: a period of transition. *Progress in Human Geography* **6**, 190–215.

LEWIS, G. M. 1966: Regional ideas and reality in the Cis-Rocky Mountain West. *Transactions, Institute of British Geographers* **38**, 135–50.

—— 1968: Levels of living in the Northeastern United States *c.* 1960: a new approach to regional geography. *Transactions, Institute of British Geographers* **45**, 11–37.

LEWIS, P. W. 1965: Three related problems in the formulation of laws in geography. *The Professional Geographer* **17**(5), 24–7.

LEWTHWAITE, G. R. 1966: Environmentalism and determinism: a search for clarification. *Annals, Association of American Geographers* **56**, 1–23.

LEY, D. 1977a: The personality of a geographical fact. *The Professional Geographer* **29**, 8–13.

—— 1977b: Social geography and the taken-for-granted world. *Transactions, Institute of British Geographers* **NS2**, 498–512.

—— 1978: Social geography and social action. In D. Ley and M. S. Samuels (eds.), *Humanistic geography: problems and prospects* (Chicago: Maaroufa Press), 41–57.

—— 1980: *Geography without man: a humanistic critique*. Research Paper 24, School of Geography, University of Oxford.

—— 1981: Behavioral geography and the philosophies of meaning. In K. R. Cox and R. G. Golledge (eds.), *Behavioral problems in geography revisited* (London: Methuen), 209–30.

LEY, D. and SAMUELS, M. S. 1978: Introduction: contexts of modern humanism in geography. In D. Ley and M. S. Samuels (eds.), *Humanistic geography: prospects and problems* (Chicago: Maaroufa Press), 1–18.

LIVINGSTONE, D. N. and HARRISON, R. T. 1981: Immanuel Kant, subjectivism, and human geography: a preliminary investigation. *Transactions, Institute of British Geographers* **NS6**, 359–74.

LÖSCH, A. 1954: *The economics of location*. New Haven, Conn.: Yale University Press.

LOWENTHAL, D. 1961: Geography, experience, and imagination: towards a geographical epistemology. *Annals, Association of American Geographers* **51**, 241–60.

—— (ed.) *George Perkins Marsh: man and nature*. Cambridge, Mass.: Harvard University Press.

—— 1968: The American scene. *Geographical Review* **48**, 61–88.

—— 1975: Past time, present place: landscape and memory. *The Geographical Review* **65**, 1–36.

LOWENTHAL, D. and BOWDEN, M. J. (eds.) 1975: *Geographies of the mind: essays in historical geosophy in honor of John Kirkland Wright*. New York: Oxford University Press.

LOWENTHAL, D. and PRINCE, H. C. 1965: English landscape tastes. *Geographical Review* **55**, 186–222.

LOWENTHAL, D. *et al.* 1973: Report of the AAG Task Force on environmental quality. *The Professional Geographer* **25**, 39–46.

LUKERMANN, F. 1958: Towards a more geographic economic geography. *The Professional Geographer* **10**(1), 2–10.

—— 1960a: On explanation, model, and prediction. *The Professional Geographer* **12**(1), 1–2.

—— 1960b: The geography of cement? *The Professional Geographer* **12**(4), 1–6.

—— 1961: The role of theory in geographical inquiry. *The Professional Geographer* **13**(2), 1–6.

—— 1965: Geography; de facto or de jure. *Journal of the Minnesota Academy of Science* **32**, 189–96.

LYNCH, K. 1960: *The image of the city*. Cambridge, Mass.: MIT Press.

McCARTY, H. H. 1940: *The geographic basis of American economic life*. New York: Harper & Brothers.

—— 1952: McCarty on McCarthy: the spatial distribution of the McCarthy vote 1952. Unpublished Paper, Department of Geography, State University of Iowa, Iowa City.

—— 1953: An approach to a theory of economic geography. *Annals, Association of American Geographers* **43**, 183–4.

—— 1954: An approach to a theory of economic geography. *Economic Geography* **30**, 95–101.

—— 1958: Science, measurement, and area analysis. *Economic Geography* **34**, facing page 283.

—— 1979: Geography at Iowa. *Annals, Association of American Geographers* **69**, 121–4.

McCARTY, H. H., HOOK, J. C. and KNOS, D. S.1956: *The measurement of association in industrial geography*. Department of Geography, State University of Iowa, Iowa City.

McCARTY, H. H. and LINDBERG, J. B. 1966: *A preface to economic geography*. Englewood Cliffs, NJ: Prentice-Hall.

McDANIEL, R. and ELIOT HURST, M. E. 1968: *A systems analytic approach to economic geography*. Commission on College Geography, Publication 8, Association of American Geographers, Washington, DC.

MACKAY, J. R. 1958: The interactance hypothesis and boundaries in Canada: a preliminary study. *The Canadian Geographer* **11**, 1–8.

McKINNEY, W. M. 1968: Carey, Spencer, and modern geography. *The Professional Geographer* **20**, 103–6.

McTAGGART, W. D. 1974: Structuralism and universalism in geography: reflections on contributions by H. C. Brookfield. *The Australian Geographer* **12**, 510–16.

MABOGUNJE, A. K. 1977: In search of spatial order: geography and the new programme of urbanization in Nigeria. In D. R. Deskins *et al.* (eds), *Geographic humanism, analysis and social action: a half century of geography at Michigan*. Michigan Geography Publications No. 17 (Ann Arbor), 347–76.

MAGEE, B. 1975: *Popper*, London: Fontana.

MANION, T. and WHITELEGG, J. 1979: Radical geography and Marxism. *Area* **11**, 122–4.

MANNERS, I. R. and MIKESELL, M. W. (eds.) 1974: *Perspectives on environment*. Commission on College Geography, Association of American Geographers, Washington.

MARCHAND, B. 1978: A dialectical approach in geography. *Geographical Analysis* **10**, 105–19.

MARCUS, M. G. 1979: Coming full circle: physical geography in the twentieth century. *Annals, Association of American geographers* **69**, 521–32.

MARTIN, A. F. 1951: The necessity for determinism. *Transactions, Institute of British Geographers* **17**, 1–12.

MARTIN, G. J. 1981: Ontography and Davisian physiography. In B. W. Blouet (ed.), *The origins of academic geography in the United States* (Hamden, Conn.: Archon Books), 279–90.

MARTIN, R. L. and OEPPEN, J. 1975: The identification of regional forecasting models using space-time correlation functions. *Transactions, Institute of British Geographers* **66**, 95–118.

MASSAM, B. H. 1976: *Location and space in social administration*. London: Edward Arnold.

MASSEY, D. 1975: Behavioral research. *Area* **7**, 201–3.

—— 1982: *The regional problem in Britain*, London: Macmillan.

MASSEY, D. and MEEGAN, R. A. 1979: The geography of industrial reorganization. *Progress in Planning* **10**, 155–237.

—— 1982: *The anatomy of job loss*. London: Methuen.

MASTERMAN, M. 1970: The nature of a paradigm. In I. Lakatos and A. Musgrave (eds.), *Criticism and the growth of knowledge* (London: Cambridge University Press), 59–90.

MAY, J. A. 1970: *Kant's concept of geography: and its relation to recent geographical thought*. Department of Geography, University of Toronto, Research Publication 4, Toronto.

—— 1972: A reply to Professor Hartshorne. *The Canadian Geographer* **16**, 79–81.

MAYER, H. M. 1954: Urban geography. In P. E. James and C. F. Jones (eds.), *American geography: inventory and prospect* (Syracuse: Syracuse University Press), 142–66.

MEAD, W. R. 1980: Regional geography. In E. H Brown (ed.), *Geography, yesterday and tomorrow* (Oxford: oxford University Press), 292–302.

MEADOWS, D. H. *et al.* 1972: *The limits to growth*. New York: Universal Books.

MEINIG, D. W. 1972: American wests: preface to a geographical introduction. *Annals, Association of American Geographers* **62**, 159–84.

MERCER, D. C. 1977: *Conflict and consensus in human Geography*. Monash Publications in Geography No. 17, Clayton, Victoria, Australia.

MERCER, D. C. and POWELL, J. M. 1972: *Phenomenology and related non-positivistic viewpoints in the social sciences*. Monash Publications in Geography, No. 1, Clayton, Victoria, Australia.

MEYER, D. R. 1972: Geographical population data: statistical description not statistical inference. *The Professional Geographer* **24**, 26–8.

MIKESELL, M. W. 1967: Geographical perspectives in anthropology. *Annals, Association of American Geographers* **57**, 617–34.

—— 1969: The borderlands of geography as a social science. In M. Sherif and C. W. Sherif (eds.), *Interdisciplinary relationships in the social sciences* (Chicago: Aldine Publishing Company), 227–48.

—— (ed.) 1973: *Geographers abroad: essays on the prospects of research in foreign areas*. Department of Geography, University of Chicago, Research Paper 152, Chicago.

—— 1974: Geography as the study of environment: an assessment of some old and new commitments. In I. R. Manners and M. W. Mikesell (eds.), *Perspectives on environment*, Commission on College Geography, Association of American Geographers (Washington), 1–23.

—— 1978: Tradition and innovation in cultural geography. *Annals, Association of American Geographers* **68**, 1–16.

—— 1981: Continuity and change. In B. W. Blouet (ed.), *The origins of academic geography in the United States* (Hamden, Conn.: Archon Books), 1–15.

MONTEFIORE, A. G. and WILLIAMS, W. M. 1955: Determinism and possibilism. *Geographical Studies* **2**, 1–11.

MOODIE, D. W. and LEHR, J. C. 1976: Fact and theory in historical geography. *The Professional Geographer* **28**, 132–6.

MORGAN, M. A. 1967: Hardware models in geography. In R. J. Chorley and P. Haggett (eds.), *Models in geography* (London: Methuen), 727–74.

MORGAN, W. B. and MOSS, R. P. 1965: Geography and ecology: the concept of the community and its relationship to environment. *Annals, Association of American Geographers* **55**, 339–50.

MORRILL, R. L. 1965: *Migration and the growth of urban settlement*. Lund Studies in Geography, Series B, 24, Lund: C. W. K. Gleerup.

—— 1968: Waves of spatial diffusion. *Journal of Regional Science* **8**, 1–18.

—— 1969: Geography and the transformation of society. *Antipode* **1**(1), 6–9.

—— 1970a: *The spatial organization of society*. Belmont California: Wadsworth, 2nd edn. 1974.

—— 1970b: Geography and the transformation of society: part II. *Antipode* **2**(1), 4–10.

—— 1974: Review of D. Harvey, *Social Justice and the City. Annals, Association of American Geographers* **64**, 475–7.

—— 1981: *Political redistricting.* Resource Publications in Geography, Association of American Geographers, Washington, DC.

MORRILL, R. L. and GARRISON, W. L. 1960: Projections of interregional patterns of trade in wheat and flour. *Economic Geography* **36**, 116–26.

MORRILL, R. L. and WOHLENBERG, E. H. 1971: *The geography of poverty in the United States.* New York: McGraw Hill.

MOSS, R. P. 1970: Authority and charisma: criteria of validity in geographical method. *South African Geographical Journal* **52**, 13–37.

—— 1977: Deductive strategies in geographical generalization. *Progress in Physical Geography* **1**, 23–39.

MOSS, R. P. and MORGAN, W. B. 1967: The concept of the community: some applications in geographical research. *Transactions, Institute of British Geographers* **41**, 21–32.

MUIR, R. 1975: *Modern political geography.* London: Macmillan.

—— 1978: Radical geography or a new orthodoxy? *Area* **10**, 322–7.

—— 1979: Radical geography and Marxism. *Area* **11**, 126–127.

MULKAY, M. J. 1975: Three models of scientific development. *Sociological Review* **23**, 509–26.

—— 1976: Methodology in the sociology of science: some reflections on the study of radio astronomy. In G. Lemaine *et al.* (eds.), *Perspectives in the emergence of scientific disciplines* (The Hague: Mouton), 207–20.

—— 1978: Consensus in science. *Social Science Information* **17**, 107–22.

MULKAY, M. J., GILBERT, G. N. and WOOLGAR, S. 1975: Problem areas and research networks in science. *Sociology* **9**, 187–203.

MULLER-WILLE, C. 1978: The forgotten heritage: Christaller's antecedents. In B. J. L. Berry (ed.), *The nature of change in geographical ideas* (de Kalb: Northern Illinois University Press), 37–64.

MURDIE, R. A. 1969: *Factorial ecology of metropolitan Toronto 1951–1961.* University of Chicago, Department of Geography, Research Paper 116, Chicago.

MYRDAL, G. 1957: *Economic theory and underdeveloped regions.* London: Duckworth.

NATIONAL ACADEMY OF SCIENCES – NATIONAL RESEARCH COUNCIL 1965: *The science of geography.* Washington: NAS-NRC.

NEFT, D. 1966: *Statistical Analysis for Areal Distributions.* Monograph 2, Regional Science Research Institute, Philadelphia.

NEWMAN, J. L. 1973: The use of the term 'hypothesis' in geography. *Annals, Association of American Geographers* **63**, 22–7.

NYSTUEN, J. D. 1963: Identification of some fundamental spatial concepts. *Papers of the Michigan Academy of Science,* Arts, and Letters, 48, 373–84. Reprinted in B. J. L. Berry and D. F. Marble (eds.), *Spatial analysis* (Englewood Cliffs, NJ: Prentice-Hall), 35–41.

ODUM, H. W. and MOORE, H. E. 1938: *American regionalism – a cultural-historical approach to national integration.* New York: H. Holt & Company.

OLSSON, G. 1965: *Distance and human interaction: a review and bibliography.* Regional Science Research Institute, Bibliography Series Number 2, Philadelphia.

—— 1969: Inference problems in locational analysis. In K. R. Cox and R. G. Golledge (eds.), *Behavioral problems in geography: a symposium*, Northwestern University Studies in Geography 17 (Evanston), 14–34.

O'RIORDAN, T. 1971a: Environmental management. In C. Board *et al.* (eds.), *Progress in Geography* 3, (London: Edward Arnold), 173–231.

—— 1971b: *Perspectives in resource management*. London: Pion.

—— 1976: *Environmentalism*. London: Pion.

PAHL, R. E. 1965: Trends in social geography. In R. J. Chorley and P. Haggett (eds.), *Frontiers in geographical teaching* (London: Methuen), 81–100.

—— 1969: Urban social theory and research. *Environment and Planning* 1, 143–54. Reprinted in R. E. Pahl 1970, *Whose City?* (London: Longman), 209–25.

—— 1975: *Whose city? and other essays*. Harmondsworth: Penguin Books (2nd edn).

PALM, R. 1979: Financial and real estate institutions in the housing market. In D. T. Herbert and R. J. Johnston (eds.), *Geography and the urban environment*, vol. 2 (London: John Wiley), 83–124.

PAPAGEORGIOU, G. J. 1969: Description of a basis necessary to the analysis of spatial systems. *Geographical Analysis* 1, 213–15.

—— (ed.) 1976: *Mathematical land use theory*. Lexington, Mass.: D. C. Heath.

PARKES, D. N. and THRIFT, N. J. 1980: *Times, Spaces and Places*. Chichester: John Wiley.

PARSONS, J. J. 1977: Geography as exploration and discovery. *Annals, Association of American Geographers* 67, 1–16.

PATERSON, J. H. 1974: Writing regional geography. In C. Board *et al.* (eds.), *Progress in Geography* 6, (London: Edward Arnold), 1–26.

PATMORE, J. A. 1970: *Land and leisure*. Newton Abbott: David & Charles.

PEACH, C. and SMITH, S. J. 1981: Introduction. In C. Peach, V. Robinson and S. J. Smith, (eds.), *Ethnic segregation in cities* (London: Croom Helm), 9–24.

PEET, J. R. 1971: Poor, hungry America. *The Professional Geographer* 23, 99–104.

—— 1975a: Inequality and poverty: a marxist-geographic theory. *Annals, Association of American Geographers* 65, 564–71.

—— 1975b: The geography of crime: a political critique. *The Professional Geographer* 27, 277–80.

—— 1976a: Further comments on the geography of crime. *The Professional Geographer* 28, 96–100.

—— 1976b: Editorial: radical geography in 1976. *Antipode* 8(3), inside cover.

—— 1977: The development of radical geography in the United States. *Progress in Human Geography* 1, 240–63.

—— (ed.) 1978: *Radical geography*. London: Methuen.

—— 1979: Societal contradiction and marxist geography. *Annals, Association of American Geographers* 69, 164–9.

—— 1980: The transition from feudalism to capitalism. In J. R. Peet (ed.), *An introduction to marxist theories of underdevelopment*. Publication HG/14, Department of Human Geography, Australian National University (Canberra), 51–74.

PEET, J. R. and LYONS, J. V. 1981: Marxism: dialectical materialism, social formation and the geographic relations. In M. E. Harvey and B. P. Holly (eds.), *Themes in geographic thought* (London: Croom Helm), 187–205.

PELTIER, L. C. 1954: Geomorphology. In P. E. James and C. F. Jones (eds.), *American geography: inventory and prospect* (Syracuse: Syracuse University Press), 362–81.

PENNING-ROWSELL, E. C. 1981: Fluctuating fortunes in gauging landscape value. *Progress in Human Geography* **5**, 25–41.

PERRY, P. J. 1969: H.C. Darby and historical geography: a survey and review. *Geographische Zeitschrift* **57**, 161–77.

—— 1979: Beyond Domesday. *Progress in Human Geography* **3**, 407–16.

PETCH, J. R. and HAINES-YOUNG, R. H. 1980: The challenge of critical rationalism for methodology in physical geography. *Progress in Physical Geography* **4**, 63–78.

PETER, L. and HULL, R. 1969: *The Peter principle.* London: Bantam Books.

PHILBRICK, A. K. 1957: Principles of areal functional organization in regional human geography. *Economic Geography* **33**, 299–366.

PIPKIN, J. S. 1981: Cognitive behavioral geography and repetitive travel. In K. R. Cox and R. G. Golledge (eds.), *Behavioral problems in geography revisited* (London: Methuen), 145–80.

PIRIE, G. H. 1976: Thoughts on revealed preferences and spatial behaviour. *Environment and Planning* A **8**, 947–55.

PITTS, F. R. 1965: A graph theoretic approach to historical geography. *The Professional Geographer* **17**(5), 15–20.

POCOCK, D. and HUDSON, R. 1978: *Images of the urban environment.* London: Macmillan.

POOLER, J. A. 1977: The origins of the spatial tradition in geography: an interpretation. *Ontario Geography* **11**, 56–83.

POPPER, K. R. 1959: *The logic of scientific discovery.* London: Hutchinson.

—— 1967: Replies to my critics. In P. A. Schipp (ed.), *The philosophy of Karl Popper*, volume 2 (La Salle, Indiana: Open Court Publishing Company), 961–97.

—— 1970: Normal science and its dangers. In I. Lakatos and A. Musgrave (eds.), *Criticism and the growth of knowledge* (London: Cambridge University Press), 51–8.

PORTEOUS, J. D. 1977: *Environment and behavior.* Reading, Mass.: Addison-Wesley.

PORTER, P. W. 1978: Geography as human ecology. *Human geography: coming of age. American Behavioral Scientist* **22**, 15–40.

PORTER, P. W. and LUKERMANN, F. 1975: The geography of utopia. In D. Lowenthal and M. J. Bowden (eds.), *Geography of the mind: essays in historical geosophy* (New York: Oxford University Press), 197–224.

POWELL, J. M. 1970: *The public lands of Australia Felix: settlement and land appraisal in Victoria 1834–1891.* Melbourne: Oxford University Press.

—— 1971: Utopia, millenium and the cooperative ideal: a behavioral matrix in the settlement process. *The Australian Geographer* **11**, 606–18.

—— 1972: *Images of Australia.* Monash University Publications in Geography No. 3, Clayton, Victoria, Australia.

—— 1977: *Mirrors of the New World: images and image-makers in the settlement process.* Folkstone: Dawson.

—— 1980a: Thomas Griffith Taylor 1880–1963. In T. W. Freeman and P. Pinchemel (eds.), *Geographers: biobibliographical studies*, volume 5 (London: Mansell), 141–54.

—— 1980b: The haunting of Saloman's house: geography and the limits of science. *Australian Geographer* **14**, 327–41.

—— 1981: Editorial comment: 'professional' geography into the eighties – ? *Australian Geographical Studies* **19**, 228–30.

PRED, A. 1965a: The concentration of high value-added manufacturing. *Economic Geography* **41**, 108–32.

—— 1965b: Industrialization, initial advantage, and American metropolitan growth. *Geographical Review* **55**, 158–85.

—— 1967: *Behavior and location: foundations for a geographic and dynamic location theory. Part I.* Lund: C. W. K. Gleerup.

—— 1969: *Behavior and location: foundations for a geographic and dynamic location theory. Part II.* Lund: C. W. K. Gleerup.

—— 1973: Urbanization, domestic planning problems and Swedish geographic research. In C. Board *et al.* (eds.), *Progress in Geography* **5**, London: Edward Arnold, 1–77.

—— 1972a: The choreography of existence: comments on Hägerstrand's time-geography and its usefulness. *Economic Geography* **53**, 207–21.

—— 1977b: *City-systems in advanced economies.* London: Hutchinson.

—— 1979: The academic past through a time-geographic looking glass. *Annals, Association of American Geographers* **69**, 175–80.

—— 1981: Production, family, and free-time projects: a time-geographic perspective on the individual and societal change in nineteenth century US cities. *Journal of Historical Geography* **7**, 3–36.

PRED, A. R. and KIBEL, B. M. 1970: An application of gaming simulation to a general model of economic locational processes. *Economic Geography* **46**, 136–56.

PRED, A. and PALM, R. 1978: The status of American women: a time-geographic view. In D. A. Lanegran and R. Palm (eds.), *An invitation to geography* (2nd edn) (New York: McGraw Hill), 99–109.

PRINCE, H. C. 1971a: Real, imagined and abstract worlds of the past. In C. Board *et al.* (eds.), *Progress in Geography* **3**, London: Edward Arnold, 1–86.

—— 1971b: America! America? Views on a pot melting 1. Questions of social relevance. *Area* **3**, 150–3.

—— 1979: About half Marx for the transition from feudalism to capitalism. *Area* **11**, 47–51.

PRUNTY, M. C. 1979: Clark in the early 1940s. *Annals, Association of American Geographers* **69**, 42–5.

QUAINI, M. 1982: *Geography and marxism.* Oxford: Blackwell.

RADFORD, J. P. 1981: The social geography of the nineteenth century US city. In D. T. Herbert and R. J. Johnston (eds.), *Geography and the urban environment*, volume 4 (Chichester: John Wiley), 257–93.

RAVENSTEIN, E. G. 1885: The laws of migration. *Journal of the Royal Statistical Society* **48**, 167–235.

RAWSTRON, E. M. 1958: Three principles of industrial location. *Transactions, Institute of British Geographers* **25**, 135–42.

RAY, D. M., VILLENEUVE, P. Y. and ROBERGE, R. A. 1974: Functional prerequisites, spatial diffusion, and allometric growth. *Economic Geography* **50**, 341–51.

REES, P. H. and WILSON, A. G. 1977: *Spatial population analysis.* London: Edward Arnold.

REISER, R. 1973: The territorial illusion and behavioural sink: critical notes on behavioural geography. *Antipode* **5**(3), 52–7.

RELPH, E. 1970: An inquiry into the relations between phenomenology and geography. *The Canadian Geographer* **14**, 193–201.

—— 1976: *Place and placelessness*. London: Pion.

—— 1977: Humanism, phenomenology, and geography, *Annals, Association of American Geographers* **67**, 177–9.

—— 1981a: Phenomenology. In M. E. Harvey and B. P. Holly (eds.), *Themes in geographic thought* (London: Croom Helm), 99–114.

—— 1981b: *Rational landscapes and humanistic geography*. London: Croom Helm.

RENFREW, A. C. 1981: Space, time and man. *Transactions, Institute of British Geographers* **NS6** 257–78.

REYNOLDS, R. B. 1956: Statistical methods in geographical research. *Geographical Review* **46**, 129–32.

RHIND, D. W. 1981: Geographical information systems in Britain. In N. Wrigley and R. J. Bennett (eds.), *Quantitative geography* (London: Routledge & Kegan Paul), 17–35.

RHIND, D. W. and ADAMS, T. A. 1980: Recent developments in surveying and mapping. In E. H. Brown (ed.), *Geography, yesterday and tomorrow* (Oxford: Oxford University Press), 181–99.

RHIND, D. W. and HUDSON, R. 1981: *Land use*, London: Methuen.

RIDDELL, J. B. 1970: *The spatial dynamics of modernization in Sierra Leone*. Evanston, Ill.: Northwestern University Press.

RIMMER, P. J. 1978: Redirections in transport geography. *Progress in Human Geography* **2**, 76–100.

ROBINSON, A. H. 1956: The necessity of weighting values in correlation analysis of area data. *Annals, Association of American Geographers* **46**, 233–6.

—— 1961: On perks and pokes. *Economic Geography* **37**, 181–3.

—— 1962: Mapping the correspondence of isarithmic maps. *Annals, Association of American Geographers* **52**, 414–25.

ROBINSON, A. H. and BRYSON, R. A. 1957: A method for describing quantitatively the correspondence of geographical distributions. *Annals, Association of American Geographers* **47**, 379–91.

ROBINSON, A. H. LINDBERG, J. B. and BRINKMAN, L. W. 1961: A correlation and regression analysis applied to rural farm densities in the Great Plains. *Annals, Association of American Geographers* **51**, 211–21.

ROBINSON, M. E. 1982: Representation, misrepresentation, and 'uncritical rhetoric'. *The Professional Geographer* **34**, 224–6.

ROBSON, B. T. 1969: *Urban Analysis*. Cambridge: Cambridge University Press.

—— 1972: The corridors of geography. *Area* **4**, 213–14.

—— 1982: Introduction. In B. T. Robson and J. Rees (eds.), *Geographical agenda for a changing world* (London: Social Science Research Council), 1–6.

RODER, W. 1961: Attitudes and knowledge on the Topeka flood plain. In G. F. White (ed.), *Papers on flood problems*, University of Chicago, Department of Geography, Research Paper 70 (Chicago), 62–83.

RODGERS, A. 1955: Changing locational patterns in the Soviet pulp and paper industries. *Annals, Association of American Geographers* **45**, 85–104.

ROSE, J. K. 1936: Corn yield and climate in the Corn Belt. *Geographical Review* **26**, 88–102.

ROTHSTEIN, J. 1958: *Communication, organization and science*. Colorado: Falcon's Wing Press.

ROWLES, G. D. 1978: Reflections on experiential field work. In D. Ley and M. S.

Samuels (eds.), *Humanistic geography: problems and prospects* (London: Croom Helm), 173–93.

RUSHTON, G. 1969: Analysis of spatial behavior by revealed space preference. *Annals, Association of American Geographers* **59**, 391–400.

—— 1979: On behavioral and perception geography. *Annals, Association of American Geographers* **69**, 463–4.

SAARINEN, T. F. 1979: Commentary–critique of Bunting-Guelke paper *Annals, Association of American Geographers* **69**, 464–8.

SACK, R. D. 1972: Geography, geometry and explanation. *Annals, Association of American Geographers* **62**, 61–78.

—— 1973a: Comment in reply. *Annals, Association of American Geographers* **63**, 568–9.

—— 1973b: A concept of physical space in geography. *Geographical Analysis* **5**, 16–34.

—— 1974a: The spatial separatist theme in geography. *Economic Geography* **50**, 1–19.

—— 1974b: Chorology and spatial analysis. *Annals, Association of American Geographers* **64**, 439–52.

—— 1980: *Conceptions of space in social thought.* London: Macmillan.

SAMUELS, M. S. 1978: Existentialism and human geography. In D. Ley and M. S. Samuels (eds.), *Humanistic geography: problems and prospects* (Chicago: Maaroufa Press), 22–40.

SANDBACH, F. 1980: *Environment, ideology and policy.* Oxford: Blackwell.

SANTOS, M. 1974: Geography, marxism and underdevelopment. *Antipode* **6**(3), 1–9.

SAUER, C. O. 1925: The morphology of landscape. *University of California Publications in Geography* **2**, 19–54.

—— 1941: Foreword to historical geography. *Annals, Association of American Geographers* **31**, 1–24.

—— 1956: The education of a geographer. *Annals, Association of American Geographers* **46**, 287–99.

SAYER, R. A. 1979: Epistemology and conceptions of people and nature in geography. *Geoforum* **10**, 19–44.

—— 1981: Defensible values in geography. In D. T. Herbert and R. J. Johnston (eds.), *Geography and the urban environment*, volume 4 (Chichester: John Wiley), 29–56.

—— 1982: Explanation in economic geography. *Progress in Human Geography* **6**, 68–88.

SCHAEFFER, F. K. 1953: Exceptionalism in geography: a methodological examination. *Annals, Association of American Geographers* **43**, 226–49.

SCOTT, A. J. 1982: The meaning and social origins of discourse on the spatial foundations of society. In P. R. Gould and G. Olsson (eds.), *A search for Common Ground* (London: Pion), 141–56.

SHANNON, G. W. and DEVER, G. E. A. 1974: *Health care delivery: spatial perspectives.* New York: McGraw Hill.

SHEPPARD, E. S. 1979: Gravity parameter estimation. *Geographical Analysis* **11**, 120–33.

SIDDALL, W. R. 1961: Two kinds of geography. *Economic Geography* **36**, facing page 189.

SIMON, H. A. 1957: *Models of man: social and rational.* New York: John Wiley.

SINCLAIR, J. G. and KISSLING, C. C. 1971: A network analysis approach to fruit distribution planning. *Proceedings, Sixth New Zealand Geography Conference* Volume 1, Christchurch, 131-6.

SLATER, D. 1973: Geography and underdevelopment – 1. *Antipode* 5(3), 21-33.

—— 1975: The poverty of modern geographical enquiry. *Pacific Viewpoint* **16**, 159-76.

SMAILES, A. E. 1946: The urban mesh of England and Wales. *Transactions and Papers, Institute of British Geographers* **11**, 85-101.

SMITH, C. T. 1965: Historical geography: current trends and prospects. In R. J. Chorley and P. Haggett (eds.), *Frontiers in geographical teaching* (London: Methuen), 118-43.

SMITH, D. M. 1971: America! America? Views on a pot melting. 2. Radical geography – the next revolution? *Area* **3**, 153-7.

—— 1973a: Alternative 'relevant' professional roles. *Area* **5**, 1-4.

—— 1973b: *The geography of social well-being in the United States.* New York: McGraw Hill.

—— 1977: *Human geography: a welfare approach.* London: Edward Arnold.

—— 1981: *Industrial location: an economic geographical analysis* (2nd edn). New York: John Wiley.

SMITH, N. 1979: Geography, science and post-positivist modes of explanation. *Progress in Human Geography* **3**, 365-83.

SMITH, W. 1949: *An economic geography of Great Britain.* London: Methuen.

SOJA, E. W. 1968: *The geography of modernization in Kenya.* Syracuse: Syracuse University Press.

—— 1980: The socio-spatial dialectic. *Annals, Association of American Geographers* **70**, 207-25.

SPATE, O. H. K. 1957: How determined is possibilism? *Geographical Studies* **4**, 3-12.

—— 1960a: Quantity and quality in geography. *Annals, Association of American Geographers* **50**, 477-94.

—— 1960b: Lord Kelvin rides again. *Economic Geography* **36**, facing page 1.

—— 1963: Letter to the editor. *Geography* **48**, 206.

SPENCER, H. 1892: *A system of synthetic philosophy, volume 1 First Principles* (4th edn). New York: Appleton.

STAMP, L. D. 1946: *The land of Britain.* London: Longman.

—— 1966: Ten years on. *Transactions, Institute of British Geographers* **40**, 11-20.

STAMP, L. D. and BEAVER, S. H. 1947: *The British Isles.* London: Longman.

STEEL, R. W. 1974: The Third World: geography in practice. *Geography* **59**, 189-207.

—— 1982: Regional geography in practice. *Geography* **67**, 2-8.

STEGMULLER, W. 1976: *The structure and dynamics of theories.* New York: Springer-Verlag.

STEWART, J. Q. 1947: Empirical mathematical rules concerning the distribution and equilibrium of population. *Geographical Review* **37**, 461-85.

—— 1956: The development of social physics. *American Journal of Physics* **18**, 239-53.

STEWART, J. Q. and WARNTZ, W. 1958: Macrogeography and social science, *Geographical Review* **48**, 167-184.

—— 1959: Physics of population distribution. *Journal of Regional Science* **1**, 99-123.

STODDART, D. R. 1965: Geography and the ecological approach: the ecosystem as a geographic principle and method. *Geography* **50**, 242-51.

—— 1966: Darwin's impact on geography. *Annals, Association of American Geographers* **56**, 683–98.

—— 1967a: Growth and structure of geography. *Transactions, Institute of British Geographers* **41**, 1–19.

—— 1967b: Organism and ecosystem as geographic models. In R. J. Chorley and P. Haggett (eds.), *Models in geography* (London: Methuen), 511–47.

—— 1975a: The RGS and the foundations of geography at Cambridge. *Geographical Journal* **141**, 216–39.

—— 1975b: Kropotkin, Reclus and 'relevant' geography. *Area* **7**, 188–90.

—— 1977: The paradigm concept and the history of geography. Abstract of a paper for the conference of the International Geographical Union Commission on the History of Geographic Thought, Edinburgh.

—— 1981a: The paradigm concept and the history of geography. In D. R. Stoddart (ed.), *Geography, ideology and social concern* (Oxford: Blackwell), 70–80.

—— 1981b: Ideas and interpretation in the history of geography. In D. R. Stoddart (ed.), *Geography, ideology and social concern* (Oxford: Blackwell) 1–7.

STOUFFER, S. A. 1940: Intervening opportunities: a theory relating mobility and distance. *American Sociological Review* **5**, 845–67.

SUPPE, F. 1977a: The search for philosophic understanding of scientific theories. In F. Suppe (ed.), *The structure of scientific theories* (Urbana: University of Illinois Press), 3–233.

—— 1977b: Exemplars, theories and disciplinary matrices. In F. Suppe (ed.), *The structure of scientific theories* (Urbana: University of Illinois Press), 473–99.

—— 1977c: Afterword – 1977. In F. Suppe (ed.), *The structure of scientific theories* (Urbana: University of Illinois Press), 617–730.

SVIATLOVSKY, E. E. and EELS, W. C. 1937: The centrographical method and regional analysis. *Geographical Review* **27**, 240–54.

TAAFFE, E. J. 1974: The spatial view in context. *Annals, Association of American Geographers* **64**, 1–16.

—— 1970: *Geography*. Englewood Cliffs, NJ: Prentice-Hall.

—— 1979: In the Chicago area. *Annals, Association of American Geographers* **69**, 133–8.

TAAFFE, E. J., MORRILL, R. L. and GOULD, P.R. 1963: Transport expansion in underdeveloped countries: a comparative analysis. *Geographical Review* **53**, 503–29.

TATHAM, G. 1953: Environmentalism and possibilism. In G. Taylor (ed.), *Geography in the twentieth century* (London Methuen), 128–64.

TAYLOR, E. G. R 1937: Whither geography? a review of some recent geographical texts. *Geographical Review* **27**, 129–35.

TAYLOR, P .J. 1976: An interpretation of the quantification debate in British geography. *Transactions, Institute of British Geographers* NS1, 129–42.

—— 1978: Political geography. *Progress in Human Geography* **2**, 53–62.

—— 1981: Factor analysis in geographical research. In R. J. Bennett (ed.), *European progress in spatial analysis* (London: Pion), 251–67.

—— 1982: A materialist framework for political geography. *Transactions, Institute of British Geographers* NS7, 15–34.

TAYLOR, P. J. and JOHNSTON, R. J. 1979: *Geography of elections*. Harmondsworth: Penguin Books.

TAYLOR, P. J. and GUDGIN, G. 1976: A statistical theory of electoral redistricting. *Environment and Planning* A **8**, 43–58.

THOMAN, R. S. 1965: Some comments on *The Science of Geography*. *The Professional Geographer* **17**(6), 8–10.

THOMAS, E. N. 1960: Areal associations between population growth and selected factors in the Chicago urbanized area. *Economic Geography* **36**, 158–70.

—— 1968: Maps of residuals from regression. In B. J. L. Berry and D. F. Marble (eds.), *Spatial analysis* (Englewood Cliffs, NJ: Prentice-Hall), 326–52.

THOMAS, E. N. and ANDERSON, D. L. 1965: Additional comments on weighting values in correlation analysis of areal data. *Annals, Association of American Geographers* **55**, 492–505.

THOMAS, R. W. 1982: *Information statistics in geography*. CATMOG 31, Geo Abstracts Ltd, Norwich.

THOMAS, W. L. (ed.) 1956: *Man's role in changing the face of the earth*. Chicago: University of Chicago Press.

THOMPSON, J. H. *et al.* 1962: Toward a geography of economic health: the case of New York state. *Annals, Association of American Geographers* **52**, 1–20.

THRIFT, N. J. 1977: *An introduction to time geography*. Concepts and Techniques in Modern Geography **13**, Norwich: Geo Abstracts Ltd.

—— 1979: Unemployment in the inner city: urban problem or structural imperative? a review of the British experience. In D. T. Herbert and R. J. Johnston (eds.), *Geography and urban environment: progress in research and applications*, volume 2 (London: John Wiley), 125–226.

—— 1981: Behavioural geography. In N. Wrigley and R. J. Bennett (eds.), *Quantitative geography* (London: Routledge & Kegan Paul), 352–65.

TIMMS, D. 1965: Quantitative techniques in urban social geography. In R. J. Chorley and P. Haggett (eds.), *Frontiers in geographical teaching* (London: Methuen), 239–65.

TOCALIS, T .R. 1978: Changing theoretical foundations of the gravity concept of human interaction. In B. J. L. Berry, (ed.), *The nature of change in geographical ideas*. (de Kalb: Northern Illinois University Press), 65–124.

TOULMIN, S. E. 1970: Does the distinction between normal and revolutionary science hold water? In I. Lakatos and A. Musgrave (eds.), *Criticism and the growth of knowledge* (London: Cambridge University Press), 39–48.

TREWARTHA, G. T. 1973: Comments on geography and public policy. *The Professional Geographer* **25**, 78–9.

TUAN, YI-FU 1971: Geography, phenomenology, and the study of human nature. *The Canadian Geographer* **15**, 181–92.

—— 1974: Space and place: humanistic perspectives. In C. Board *et al.* (eds.), *Progress in Geography* **6**, London: Edward Arnold, 211–52.

—— 1975a: Images and mental maps. *Annals, Association of American Geographers* **65**, 205–13.

—— 1975b: Place: an experiential perspective *Geographical Review* **65**, 151–65.

—— 1976: Humanistic geography. *Annals, Association of American Geographers* **66**, 266–76.

—— 1977: *Space and place*. London: Edward Arnold.

—— 1978: Literature and geography: implications for geographical research. In D. Ley and M. S. Samuels (eds.), *Humanistic geography: prospects and problems* (Chicago: Maaroufa Press), 194–206.

TULLOCK, G. 1976: *The vote motive*. London: Institute of Economic Affairs.

ULLMAN, E. L. 1941: A theory of location for cities. *American Journal of Sociology* **46**, 853–64.

—— 1953: Human geography and area research. *Annals, Association of American Geographers* **43**, 54–66.

—— 1956: The role of transportation and the bases for interaction. In W. L. Thomas (ed.), *Man's role in changing the face of the earth* (Chicago: University of Chicago Press), 862–80.

URLICH, D. U. 1972: Migrations of the North Island Maoris 1800–1840: a systems view of migration. *New Zealand Geographer* **28**, 23–35.

URLICH CLOHER, D. 1975: A perspective on Australian urbanization. In J. M. Powell and M. Williams (eds.), *Australian space, Australian time: geographical perspectives* (Melbourne: Oxford University Press), 104–59.

VAN DEN DAELE, W. and WEINGART, P. 1976: Resistance and receptivity of science to external direction: the emergence of new disciplines under the impact of science policy. In G. Lemaine *et al.* (eds.), *Perspectives on the emergence of scientific disciplines* (The Hague: Mouton), 247–75.

VAN PAASSEN, C. 1981: The philosophy of geography: from Vidal to Hägerstrand. In A. Pred and G. Tornquist (eds.), *Space and time in geography* (Lund: C. W. K. Gleerup), 17–29.

VANCE, J. E. 1970: *The merchant's world.* Englewood Cliffs, NJ: Prentice-Hall.

—— 1978: Geography and the study of cities. *Human Geography: Coming of Age. American Behavioral Scientist* **22**, 131–49.

WAGNER, P. L. 1976: Reflections on a radical geography. *Antipode* **8**(3), 83–5.

WALKER, R. A. 1981a: A theory of suburbanization. In M. J. Dear and A. J. Scott (eds.), *Urbanization and urban planning in capitalist societies* (London: Methuen), 383–430.

—— 1981b: Left-wing libertarianism, an academic disorder: a reply to David Sibley. *The Professional Geographer* **33**, 5–9.

WALKER, R. A. and STORPER, M. 1981: Capital and industrial location. *Progress in Human Geography* **5**, 473–509.

WALMSLEY, D. J. 1972: *Systems theory: a framework for human geographical enquiry.* Research School of Pacific Studies. Department of Human Geography Publication HG/7, Australian National University, Canberra.

—— 1974: Positivism and phenomenology in human geography. *The Canadian Geographer* **18**, 95–107.

WALMSLEY, D. J. and SORENSEN, A. D. 1980: What marx for the radicals? an Antipodean viewpoint. *Area* **12**, 137–41.

WARD, D. 1971: *Cities and immigrants: a geography of change in nineteenth-century America.* New York: Oxford University Press.

WARNTZ, W. 1959a: *Toward a geography of price.* Philadelphia: University of Pennsylvania Press.

—— 1959b: Geography at mid-twentieth century. *World Politics* **11**, 442–54.

—— 1959c: Progress in economic geography. In P. E. James (ed.), *New viewpoints in geography* (Washington: National Council for the Social Studies), 54–75.

—— 1968: Letter to the editor. *The Professional Geographer* **20**, 357.

WATKINS, J. W. N. 1970: Against 'normal science'. In I. Lakatos and A. Musgrave (eds.), *Criticism and the growth of knowledge* (London: Cambridge University Press), 25–38.

WATSON, J. D. 1968: *The double helix: a personal account of the discovery of the structure of DNA.* London: Weidenfeld & Nicolson.

WATSON, J. W. 1953: The sociological aspects of geography. In G. Taylor (ed.), *Geography in the twentieth century* (London: Methuen), 453–99.

—— 1955: Geography: a discipline in distance. *Scottish Geographical Magazine* **71**, 1–13.

WATTS, S. J. and WATTS, S. J. 1978: The idealist alternative in geography and history. *The Professional Geographer* **30**, 123–7.

WEAVER, J. C. 1943: Climatic relations of American barley production. *Geographical Review* **33**, 569–88.

—— 1954: Crop-combination regions in the Middle West. *Geographical Review* **44**, 175–200.

WEBBER, M. J. 1972: *Impact of uncertainty on location.* Canberra: Australian National University Press.

—— 1977: Pedagogy again: what is entropy? *Annals, Association of American Geographers* **67**, 254–66.

WESTERN, J. S. 1978: Knowing one's place: 'the Coloured people' and the Group Areas Act in Cape Town. In D. Ley and M. S. Samuels (eds.), *Humanistic geography: problems and prospects* (Chicago: Maaroufa Press), 297–318.

WHEELER, P. B. 1982: Revolutions, research programmes and human geography. *Area* **14**, 1–6.

WHITE, G. F. 1945: *Human adjustment to floods.* University of Chicago, Department of Geography, Research Paper 29, Chicago.

—— 1972: Geography and public policy. *The Professional Geographer* **24**, 101–4.

—— 1973: Natural hazards research. In R. J. Chorley (ed.), *Directions in geography* (London: Methuen), 193–216.

WHITEHAND, J. W. R. 1970: Innovation diffusion in an academic discipline: the case of the 'new' geography. *Area* **2**(3), 19–30.

WHITEHAND, J. W. R. and PATTEN, J. H. C. (eds.) 1977: *Change in the town.* Transactions, Institute of British Geographers **2**(3).

WILLIAMS, P. R. 1978: Urban managerialism: a concept of relevance? *Area* **10**, 236–40.

—— 1982: Restructuring urban managerialism: towards a political economy of urban allocation. *Environment and Planning A* **14**, 95–106.

WILSON, A. G. 1967: A statistical theory of spatial distribution models. *Transportation Research* **1**, 253–69.

—— 1970: *Entropy in urban and regional modelling.* London: Pion.

—— 1974: *Urban and regional models in geography and planning.* London: John Wiley.

—— 1976a: Catastrophe theory and urban modelling: an application to modal choice. *Environment and Planning A* **8**, 351–46.

—— 1976b: Retailers' profits and consumers' welfare in a spatial interaction shopping model. In I. Masser (ed.), *Theory and practice in regional science* (London: Pion), 42–57.

—— 1978: Mathematical education for geographers. Department of Geography, University of Leeds, *Disscussion Paper* 211, Leeds.

—— 1981a: *Catastrophe theory and bifurcation: applications to urban and regional systems.* London: Croom Helm.

—— 1981b: *Geography and the environment: systems analytical methods.* Chichester: John Wiley.

WILSON, A. G., REES, P. H. and LEIGH, C. 1977: *Models of cities and regions.* London: John Wiley.

WISE, M. J. 1975: A university teacher of geography. *Transactions, Institute of British Geographers* **66**, 1–16.

—— 1977: On progress and geography. *Progress in Human Geography* **1**, 1–11.

WISNER, B. 1970: Introduction: on radical methodology. *Antipode* **2**, 1–3.

WOLDENBERG, M. J. and BERRY, B. J. L. 1967: Rivers and central places: analogous systems? *Journal of Regional Science* **7**, 129–40.

WOLF, L. G. 1976: Comments on the Harries-Peet controversy. *The Professional Geographer* **28**, 196–8.

WOLPERT, J. 1964: The decision process in spatial context. *Annals, Association of American Geographers* **54**, 337–58.

—— 1965: Behavioral aspects of the decision to migrate. *Papers and Proceedings, Regional Science Association* **15**, 159–72.

—— 1967: Distance and directional bias in inter-urban migratory streams. *Annals, Association of American Geographers* **57**, 605–16.

—— 1970: Departures from the usual environment in locational analysis. *Annals, Association of American Geographers* **60**, 220–9.

WOLPERT, J., DEAR, M. and CRAWFORD, R. 1975: Satellite mental health facilities. *Annals, Association of American Geographers* **65**, 24–35.

WOOLDRIDGE, S. W. 1956: *The geographer as scientist.* London: Thomas Nelson.

WOOLDRIDGE, S. W. and EAST, W. G. 1958: *The spirit and purpose of geography.* London: Hutchinson.

WOOLGAR, S. W. 1976: The identification and definition of scientific collectivities. In G. Lemaine *et al.* (eds.), *Perspectives on the emergence of scientific disciplines* (The Hague: Mouton), 233–45.

WRIGHT, J. K. 1925: *The geographical lore of the time of the crusades: a study in the history of medieval science and tradition in western Europe.* New York: American Geographical Society.

—— 1947: Terrae incognitae: the place of imagination in geography. *Annals, Association of American Geographers* **37**, 1–15.

WRIGLEY, E. A. 1965: Changes in the philosophy of geography. In R. J. Chorley and P. Haggett (eds.), *Frontiers in geographical teaching* (London: Methuen), 3–24.

WRIGLEY, N. 1979: Developments in the statistical analysis of categorical data. *Progress in Human Geography* **3**, 315–55

WRIGLEY, N. and BENNETT, R. J. (eds.) 1981: *Quantitative geography: a British view.* London: Routledge & Kegan Paul.

YOUNG, R. H. and PETCH, J. R. 1978: *The methodological limitations of Kuhn's model of science.* University of Salford, Department of Geography, Discussion Paper 8.

ZELINSKY, W. 1970: Beyond the exponentials: the role of geography in the great transition. *Economic Geography* **46**, 499–535.

—— 1978: Introduction. *Human Geography: Coming of Age. American Behavioral Scientist* **22**, 5–13.

—— 1973a: Women in geography: a brief factual report. *The Professional Geographer* **25**, 151–65.

—— 1973b: *The cultural geography of the United States.* Englewood Cliffs, NJ: Prentice-Hall.

—— 1974: Selfward bound? Personal preference patterns and the changing map of American society. *Economic Geography* **50**, 144–79.

—— 1975: The demigod's dilemma. *Annals, Association of American Geographers* **65**, 123–43.

ZIPF, G. K. 1949: *Human behavior and the principle of least effort.* New York: Hafner.

General index

academic life 9–17
agricultural geography 45, 46, 61
American Geographical Society 40, 140
anarchy 26, 217
anthropology 90, 178–9
Antipode 175–6
applied geography 58, 104, 117, 128, 156, 208, 214
archaeology 29, 96
archive 29–30
areal differentiation 55, 95
Association of American Geographers 160, 161, 162, 163, 164, 173, 207, 208
autocorrelation 105–6

behavioural geography 125–44, 148–50, 155, 157, 179, 184–6, 214
behavioural matrix 137–7
branching 26–7, 215

career structure 10–12, 207, 210–11
cartography 61–2, 66
catastrophe theory 106, 197
central place theory 63–6, 100, 114–15, 122, 125, 146
Chicago school 81, 85–6, 102–3
critical theory 183
cultural geography 79–80, 90–2, 140–4, 182

decision-making 126–31, 141, 143
diffusion 79–80, 90, 96, 100, 106, 113, 130, 139, 188
disciplinary matrix 23, 209, 218
discourse 29–30, 218

economic geography 60–1, 64–6, 76–8, 80, 86

ecosystem 115–16
educational system 35–6, 39–40, 207–8
empiricism 49, 188
entropy 118–21
environmental determinism 41–2, 48, 75
environmental images 132–3, 143–4, 154
environmentalism 173–5
episteme 30, 219
exceptionalism 52–7
exemplar 23
existentialism 152, 155, 181
external environment 30–6, 205–11

feedback 114–16
forecasting 106–7

generational model 209–11
general systems theory 121–2
geometry 65–6, 98–100, 107–9
geosophy 146
gravity model 102–4, 109

hermeneutics 113
historical geography 46–7, 55, 85, 89–91, 140–4, 148–9, 151
history 53, 56, 145
human ecology 48, 90, 116
humanistic geography 140, 144–57, 166, 182–5, 198, 216
hypothesis 71–4, 87, 120, 126, 140

idealism 90, 147–50, 189
inductive method 126–44
information theory 118–21
Institute of British Geographers 162, 164, 166, 208–9

Index to authors